Space-Time Array Communications

Vector Channel Estimation and Reception

Space-Time Array Communications

Vector Channel Estimation and Reception

Jason W. P. Ng

British Telecommunications Research Labs, UK

ICP

Imperial College Press

Published by

Imperial College Press
57 Shelton Street
Covent Garden
London WC2H 9HE

Distributed by

World Scientific Publishing Co. Pte. Ltd.
5 Toh Tuck Link, Singapore 596224
USA office: 27 Warren Street, Suite 401-402, Hackensack, NJ 07601
UK office: 57 Shelton Street, Covent Garden, London WC2H 9HE

British Library Cataloguing-in-Publication Data
A catalogue record for this book is available from the British Library.

SPACE-TIME ARRAY COMMUNICATIONS
Vector Channel Estimation and Reception

ISBN-13 978-1-86094-672-1
ISBN-10 1-86094-672-0

Printed in Singapore.

To my wife Janice
and my son Vincent

Preface

Space-time array communications – vector channel estimation and reception is stemmed from the integration of two emerging technologies for the future wireless mobile systems: space-time array processing and spread spectrum multiple access communications. The vector channel constitutes the spatial-temporal multipath propagation environment resulting from multiple reflection, refraction, diffraction and scattering phenomena experienced by the propagating signal. By applying space-time processing in spread spectrum multiple access system, the multiuser vector channel can be exploited, in accordance to its environmental context, to provide a mitigation of the various associated channel impairments. As a consequence, new dimensional layer of interference cancellation methodology can be devised, and additional form of inherent diversity can be derived, based on specifically developed vector channel estimation technique integrated at the front-end of the reception process. This book will look at the following three original integrated vector channel estimation and reception algorithms, with each dealing with different practical environmental scenario.

First is the Polarisation-Space-Time estimation and reception. Traditional array processing techniques often ignore the polarisation aspect of the signal, disregarding the fact that, in typical mobile environment, the received signal rarely takes on its transmitted polarisation. The diversity in the signal polarisation is not insignificant due to the depolarisation mechanism intrinsical in the propagation channel, and considering the recent development of portable handheld terminals which are often randomly orientated. Hence, instead of taking the polarisation factor as part of signal fading, the polarisation information inherent in the signal can be exploited to improve the accuracy and resolution of the estimation process and enhance the detection capability of the receiver.

Second is the Diffuse-Space-Time estimation and reception. Most array processing models frequently assume a rather simplistic view of the multipath propagation environment, which is made up of multiple point sources in the channel. However in typical wireless urban or suburban setup, the signal transmitted into the channel may suffer localised scattering which will inevitably create diffusion in its signal components. This consequently leads to a performance degradation when conventional array processing technique, founded on point sources assumption, is employed. A generalised diffusion framework is therefore devised to cope with both point and/or diffuse sources resulted in the signal environment. Its underlying architecture also permits ease in an extension to a co-code environmental scenario.

Third is the Doppler-Space-Time estimation and reception. Doppler spread, due to the relative motion in the environment, is detrimental in typical Multiple Input Multiple Output (MIMO) system. This can however be incorporated as a source of diversity by exploiting these multipath Doppler components in the receiver design. Unlike many MIMO works which are based on multiple independent antenna elements, the proposed MIMO array receiver is built on antenna array technology which is commonly used in applications such as the well-known smart-antenna system. It does not require the imposition of the knowledge of the channel or the need of any power control as in its conventional counterparts.

To gain a deeper understanding in the above algorithmic frameworks, other representative examples are also included at the end of the text to allow new insights into the application of these algorithms. Following that, some of the key points in the current research are highlighted, which offer some indicative views into the various potential areas for further investigative studies.

Note that this book is meant to provide a holistic picture in the treatment of the subject, bringing and linking together a number of the author's main relevant publications. It is hopeful that the readers will find the text useful, in not only gaining a comprehensive understanding but also in extending the work to other fields of research.

Jason W.P. Ng

Contents

Chapter 3: Polarisation-Space-Time Estimation and Reception 37

Chapter 4: Diffuse-Space-Time Estimation and Reception 69

CHAPTER 1

Introduction

Wireless communications technology has come a long way [1,2], dating back to the pioneering work by the Italian physicist Guglielmo Marconi (1874-1937) in the 1900s. He had demonstrated the feasibility of sending radio message signal over great distances which consequently led to the commercial realisation of the wireless telegraph stations network. It took another few decades or so before the cellular system concept was subsequently been introduced with several of its key technologies (including the seven-cell frequency-reuse technique) developed at AT&T Bell Laboratories during the 1960s [2]. This hence gave rise to the first generation (1G) analogue cellular system which was based on the Frequency Division Multiple Access (FDMA) technique. The cellular system was however made popular using the design based upon a high-capacity analogue standard developed by AT&T known as the Advanced Mobile Phone Service (AMPS). The standard was widely deployed in the United States and had gained widespread acceptance. Other popular standards [2] include the Nordic Mobile Telephone (NMT) and the Total Access Communications System (TACS) in Europe; the Nippon Telegraph and Telephone (NTT) and the Japanese

TACS (JTACS) in Japan; and the C-450 in Germany, bringing solely wireless voice access services to the Public Switched Telephone Network (PSTN).

In contrast to the first generation analogue systems, the second generation (2G) systems are designed for digital modulation transmissions. Unlike its analogue counterparts, the digital systems are able to provide flexible, secure, better quality, and higher capacity services. Advanced signal processing techniques can also be applied to the received digital signal to mitigate the effect of noise and multipath interferences encountered in the channel. The former FDMA scheme is however not implemented in the 2G systems. The scheme lacks the ability to utilise its resources efficiently as it requires the allocation of an entire frequency channel for the connection interval between the two users even though they do not converse 100% of the time in that whole duration. Digital systems which are capable of handling discontinuous transmission/reception thus pave the way for better multiple-access candidates, such as the Time Division Multiple Access (TDMA) and the Code Division Multiple Access (CDMA) schemes, to be employed in the system [2]. By far, the most successful 2G system is the Global System for Mobile communications (GSM), which is the first digital standard developed to establish cellular compatibility throughout Europe. Its success has spread to many countries, creating a mass market for mobile communications and a high terminal penetration in the global markets. Similar standards that are based upon TDMA include the North America IS-54 system which is basically a digital variation of AMPS (also known as D-AMPS), the Japanese Personal Digital Cellular (PDC) system, and etc [3]. IS-95 on the other hand is based on CDMA, which consequently forms the basis for the third generation (3G) cellular system.

Just when the second generation systems were widely being deployed, works were already underway to develop the third generation (3G) system. The projected growth in the demand for capacities, data rates and services, calls for the need to have a new network standard. The International Telecommunication Union (ITU) hence formulated an initiative called the Future Public Land Mobile Telephone Systems (FPLMTS), which was later renamed as the International Mobile

Telecommunications 2000 (IMT-2000) in 1995. The IMT-2000 is the 3G cellular standard [4] aimed at providing ubiquitous wireless access to the global telecommunication infrastructure through both satellite and terrestrial systems. Its main goals encompass a maximum data rate of 2Mbits/s, a higher capacity, and an ability to support multimedia voice/video/data services [5]. CDMA scheme, as seen employed in the second generation system, is envisaged to play an important role in its air interface protocol. In fact, the European Telecommunications Standards Institute (ETSI) has developed the Universal Mobile Telecommunications System (UMTS) based on Wideband Code Division Multiple Access (WCDMA), which is one of the major 3G mobile communications systems within the IMT-2000 framework. However to sustain the IMT-2000 vision in the long run, new enabling technology has to be included to enhance the capability of the system [6]. The use of antenna arrays, which is regarded as a core component in the future-generation mobile networks, thus becomes a viable option to be embedded in the system [6-8]. In fact, the current UMTS standard has already provided for the use of sensor array [9] and there is presently a major thrust to make space-time processing an important part of the 3G networks. Subsequent sections will hence give further insights into these two key technologies: CDMA and space-time array processing, which when combined together will play a significant role in the future of the next-generation wireless communications system.

1.1 Spread Spectrum Multiple Access Scheme

Multiple access scheme is an allocation strategy in apportioning the available physical communication resources among several simultaneous transmitters sharing a single common channel. To avoid co-channel interference (CCI), the physical resources readily accessible to allocation can be categorised in terms of its space, time or frequency domain as illustrated in Figure 1.1. As seen from the figure, Frequency Division Multiple Access - FDMA (see Figure 1.1a) ensures its resulting spectra will not overlap with one another by allocating different carrier frequency to each of the users sharing the same space and time domains;

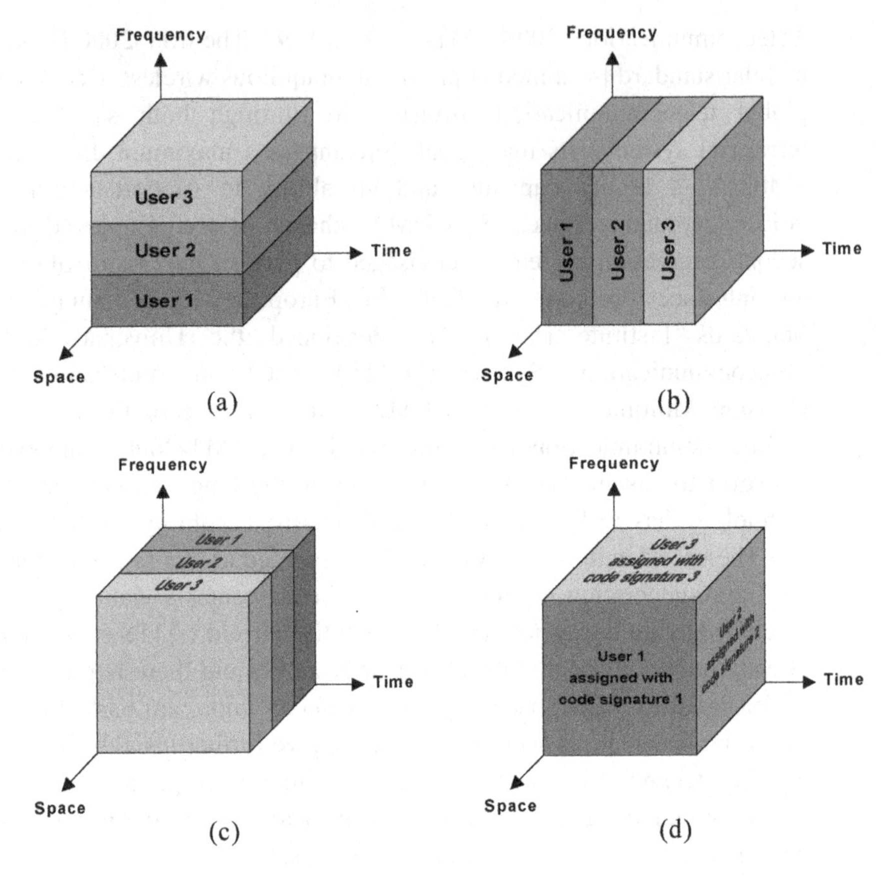

Figure 1.1: Multiple access schemes: (a) Frequency Division Multiple Access (FDMA), (b) Time Division Multiple Access (TDMA), (c) Space Division Multiple Access (SDMA), and (d) Code Division Multiple Access (CDMA).

Time Division Multiple Access - TDMA (see Figure 1.1b) on the other hand makes the channel orthogonal in the temporal domain by assigning different time slot to every of its users; and Spatial Division Multiple Access - SDMA (see Figure 1.1c) provides virtual spatial channel among the users by controlling its space domain using antenna arrays [10-12]. However, an alternative allocation strategy can be established without having to perform segregations in the space, time or frequency domain: Code Division Multiple Access - CDMA (see Figure 1.1d) allows the simultaneous usage of all the three overlapping physical resources by introducing a unique identification code assigned to each user. Unlike the above physical resources, such purposeful introduction opens up an extra domain of separation to differentiate every of its individual users.

CDMA [13] is a multiple access spread spectrum scheme originated from past military applications. The primary objectives of the scheme encompass hiding its transmitted signal from unwanted eavesdroppers (i.e. low probability of detection), encrypting its data information from unintended interceptors (i.e. low probability of interception), and immunising its communication link against deliberate jammers' interferences (i.e. anti-jamming). It involves spreading the information signal over a much greater spectrum bandwidth so that it appears virtually indistinguishable from the inherent background noise. The technological concept was then borrowed and commercialised by Qualcomm Inc. for the burgeoning cellular telecommunication industry. Since then, several variants of the CDMA schemes have been devised. The three commonly-used schemes [14] are DS-CDMA (Direct Sequence), FH-CDMA (Frequency Hopping), and TH-CDMA (Time Hopping). And the hybrids of these three schemes include DS/TH, DS/FH, FH/TH, and others [15-18]. However among these, the DS-CDMA scheme is the most widely deployed in commercial spread spectrum mobile system. In fact, the scheme, which is the focus of this book, has already been employed in the 2G IS-95 cellular network and chosen as the air-interface for the 3G UMTS standard [19].

To satisfy the requirements of the 3G mobile systems, the spread spectrum scheme possesses characteristics that have distinct advantages over other multiple access techniques. Chief among these is the universal frequency-reuse attribute that allows the users to share the same frequency band in a cellular network. Not only does this helps in increasing the efficiency of the spectrum usage, but also saves the hassle of frequency planning for the neighbouring cells or users. Besides that, it permits soft handover across multiple cell base stations by allowing connection to the new cell to be made before connection to the current cell is broken. This thus avoids the disruptive abrupt transitions encountered in conventional hard handover, consequently aids in reducing its associated dropped-call probability and improving its corresponding cell-boundary performance. Another major benefit of the scheme is the dynamic sharing of channel resources. The scheme can accommodate a higher system capacity which is dependent on the number of simultaneous active users in the cell. Unlike other multiple access schemes, its system capacity can be increased with trade-off against the quality of service (QoS). Such dynamic sharing property provides operational flexibility in its provision of services. A wide range of multimedia services with different quality requirements - such as voice, video or data service, as well as a combination of these services - can thus be supported concurrently in the system. Lastly the scheme, originated previously from military purposes, inherits the advantage of improved security and privacy. Its working function also allows coexistence operation with other narrowband microwave system.

However the spread spectrum scheme entails some disadvantages. As with other multiple access techniques involving orthogonal space, time or frequency allocation, the scheme maintains its orthogonality by means of a set of spread spectrum code signatures in a synchronous environment. But in an asynchronous setup with uncoordinated transmissions from the multiplicity of users, its non-alignment in time epochs will create quasi-orthogonality among the users, thus resulting in Multiple-Access Interference (MAI). Such level of MAI will increase with the number of users and, if left uncontrolled, may have an adverse effect on the system performance. Another drawback in the scheme is the near-far effect. If all the users in the spread spectrum scheme are transmitting at the same power level, the received signal from users who are closer to the receiver will be stronger than those who are farther away. This will inevitably lead to performance inconsistency depending on the users' location in the cell. But what is critical is that every of these nearby users, who are essentially more powerful at the receiver, will each cause some interferences to those distant users, hence creating a much higher MAI effect. This phenomenon is known as the near-far problem [14]. Power control is hence introduced in order to curb the transmitted power of those who are closer and heighten that of those who are farther to alleviate this effect. Nonetheless in practice, such power control, be it open loop or closed loop, requires knowledge of the propagation channel between the transmitter and receiver. This will consequently impose additional complexity in the system. Alternatively, appropriate design of the receiver, taking into account the above two drawbacks, can robustify the system against these shortcomings. Such robustification design in the receiver will be delineated in this study.

1.2 Antenna Array Communications

Antenna array processing for wireless communications is an evolution from the more traditional array signal processing technique. Typical technique centred primarily around conventional beamforming-based methods from a spatial perspective [20], such as the classical Bartlett beamformer [21] and Capon's Minimum Variance Distortionless Response (MVDR) beamformer [22]. Most of these techniques, however, have some fundamental limitations in their spectral resolution. The high resolution subspace-based approach, due to the introduction of the well-known MUltiple SIgnal Classification (MUSIC) algorithm [23], hence draws tremendous interest. Since then, several variants of the algorithm have been devised, such as Root-MUSIC [24], Cyclic-MUSIC [25], Beamspace-MUSIC [26], etc. Other notable subspace-based algorithms include the Estimation of Signal Parameters via Rotational Invariance Technique (ESPRIT) introduced by Roy and Kailath [27], the Minimum-Norm (Min-Norm) algorithm suggested by Kumaresan and Tufts [28], and the Weighted Subspace Fitting (WSF) method proposed by Viberg and Ottersten [29]. Another commonly-used optimal array processing model is the maximum likelihood [30] approach, encompassing Stochastic or Unconditional Maximum Likelihood (SML or UML) technique [31,32] and Deterministic or Conditional Maximum Likelihood (DML or CML) technique [33-35].

Array communications, in essence, are based from these traditional processing techniques but are extended to exploit the rich structure of the propagating signals and incorporate them within the wireless communications process. However the realisable processing gain is largely dependent on the several parameters associated with the wireless multipath propagation channel. Hence an understanding of the various channel aspects influencing the space-time processing performance of the communications system is therefore necessary.

1.2.1 Wireless Multipath Channel Characterisation

In the study of wireless communications system, the classical additive white Gaussian noise (AWGN) channel is normally the starting point in understanding the basic performance relationships. The primary source of performance degradation for such an ideal channel is solely the noise inherent in the system. However, external interferences as a result of channel propagation often have a much more significant deteriorating effect on the system performance. In practical scenarios, the transmitted signal propagating through the channel interacts with the environment in a rather complex manner. The four underlying interacting mechanisms in the environment can be categorised as follows: reflection of signal from sizeable obstruction, refraction of signal through penetrable obstruction, diffraction of signal around narrow-edged obstruction, and scattering of signal from rough-surfaced obstruction. Such complex interaction causes

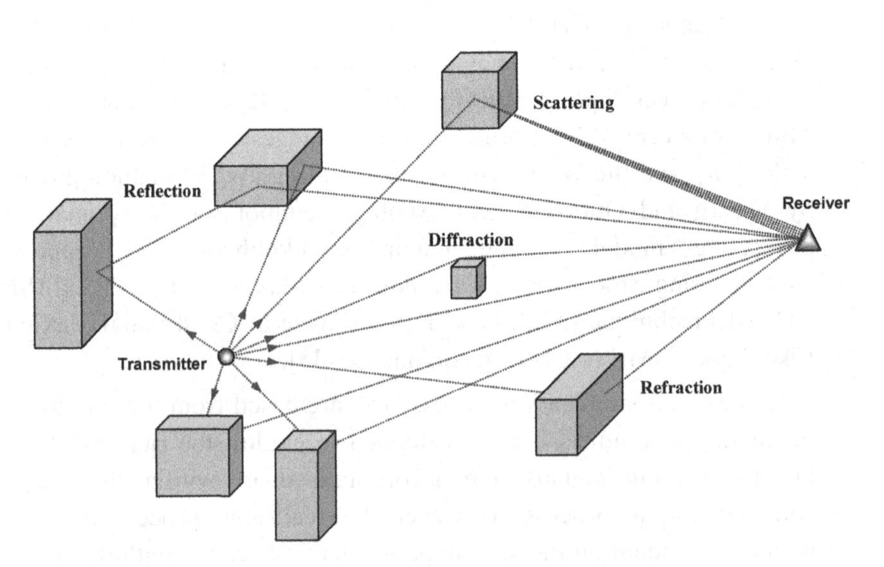

Figure 1.2: Multipath propagation environment.

multiple replicas of the signal arriving along a number of different paths at the receiver, referred to as multipath propagation. A simplified illustration of the multipath environment is as depicted in Figure 1.2.

In addition to that, the received signal also suffers a weaker power level than its original transmitted signal due to the effects of path loss and fading existing in the propagation channel. The mean path loss describes the attenuation of the signal in a free-space propagation environment, which is defined as an ideal obstruction-free transmission medium. In such free-space environment, the signal's attenuation is modelled as a function of its propagation distance and it behaves according to an inverse square law (that is, with an attenuation power exponent of 2). But in a real practical environment with the presence of obstructions, the attenuation power exponent may vary with typical values in the order of between 2 to 5 [36]. Fading, on the other hand, refers to the fluctuation in the received signal level experienced in the channel. The causes of fading can be attributed and broadly classified into two main classes: slow fading and fast fading. Slow fading, or long-term fading, is as a result of the blocking effect, also known as the shadowing effect, caused by prominent terrain contours such as hills, forests, buildings etc. Such random shadowing effect encountered by the signal along its propagation path describes a log-normal distribution about its distance-dependent mean path loss. This phenomenon is often termed as log-normal shadowing. Fast fading, or short-term fading, in contrast is due to the different multipath signals being added up with random phases, constructively or destructively, at the receiver. This hence gives rise to the rapid and dramatic fluctuations in the received signal level with its local average signal level following that of the long-term fading model. If the number of multipath is large and there is non line-of-sight (NLOS) signal component, the fast fading envelope of the received signal can be statistically approximated by a Rayleigh density function. Whereas in the presence of a direct line-of-sight (LOS) signal path, this envelope becomes no longer Rayleigh distributed and it is well described by a Ricean density function. But for some environments, experimental measurements have shown to support a better model of the fading envelope which is that of a Nakagami distribution [37].

Besides the above detrimental effects, multipath propagation also creates dispersion in the channel across a number of domains. Such channel dispersion is as a result of the received signal energy being spread in any of the frequency, time, space or polarisation dimension. The characteristics of these four different spreads - known as the Doppler spread, delay spread, angular spread and polarisation spread respectively - are briefly described as follows.

> ***Doppler spread*** describes the time-varying nature of the channel that is caused by either the relative motion between the transmitter and the receiver or by the movement of the obstructions in the channel. It is a measure of the spectral broadening width defined as the range of Doppler shift frequencies over which the received Doppler power spectrum is essentially non-zero [38]. By considering its Fourier transform pair, the reciprocal of this spectral broadening width is effectively the coherence time of the channel. Coherence time is therefore the time domain dual of Doppler spread and it represents the time duration over which the channel impulse response is essentially time-invariant. Hence a fast varying channel will tend to have a smaller coherence time, or equivalently, a larger Doppler spread. If the reciprocal bandwidth of the baseband message signal is greater than the coherence time of the channel, its channel characteristic will consequently be changed in the duration of the message signal, thus giving rise to the so-called time-selective fading situation.
>
> ***Delay spread*** is as a result of multipath arriving at the receiver with different time delay, and is defined as the propagation time difference between its longest and shortest received signal path. Similarly, its analogous Fourier transformation relationship characterises the inverse of the delay spread as a measure of the coherence

bandwidth of the channel. This time, the channel coherence bandwidth is the duality of the delay spread in the frequency domain. It is a statistical measure of the range of frequencies over which the channel passes all its spectral components with almost equal gain and linear phase. In other words, the coherence bandwidth specifies the maximum range of frequencies over which its channel response will remain relatively constant. Thus, a signal is said to experience flat fading (or frequency-flat fading) if its bandwidth is below the coherence bandwidth of the channel. In contrast, if the bandwidth of the message signal exceeds the coherence bandwidth, the time-dispersive channel is termed to exhibit frequency-selective fading. Besides that, this condition also induces another form of fading degradation which is caused by the Inter-Symbol Interference (ISI) contributions from its received multipath components.

Angular spread arises from the different arriving angles of the incoming multipath impinging on the antenna array. The amount of the spread is dependent on the signal propagation environment, which is high in an urban case and low in a rural scenario. In a similar manner, its associated coherence distance is inversely related to the angular spread, and is defined as the range of space for which their channel responses are strongly correlated. In other words, coherence distance represents the maximum spatial separation between the antenna elements whereby their fading effects are relatively the same. Thus if the angular spread is considerably large, all the incoming multipath components will be added up randomly at each of the antenna element, bringing about space-selective fading. Hence antenna array elements that are spaced above the coherence distance tend to experience uncorrelated fading which is a prerequisite for the application of space diversity [39,40].

Polarisation spread refers to the diverseness in the polarisation of the multipath electromagnetic waves incident onto the receiving antenna element. The polarisation of the received multipath component is oftentimes not what is originally intended at the antenna end of the transmitting device. This is due to the change in the polarisation state of the electromagnetic wave, brought about by the complex interaction of the wave, in the form of reflection, refraction, diffraction and scattering, within the propagation channel. Such alteration in polarisation through multipath propagation is known as depolarisation [41]. In addition to that, the random orientation of the portable transmitting device also plays a significant part in varying its intended transmitted polarisation from one state to another. This diverseness in the received polarisation state is consequently the cause of polarisation fading [42], whereby the signal fades up and down depending on the relative alignment of its electric field with respect to the receiving antenna element. Not only that, it is also the decorrelation mechanism behind the source of polarisation diversity [43,44] in order to achieve low correlation [45] between its diversity branches.

1.2.2 Space-Time Array Processing

The above degradation effects introduced by the propagation channel cause the transmitted signal to be distorted at the received end of the communication link. The general approach to improve the system performance is to employ some form of mitigation to remove or reduce the amount of such signal distortion. This is incisively the motivation underlying space-time array processing, which exploits the spatial and temporal dimensions of the received signal to overcome the various channel impairments. The technique, in principle, combines the benefits of both the time-only and the space-only processing configurations to enhance the performance of the communication link.

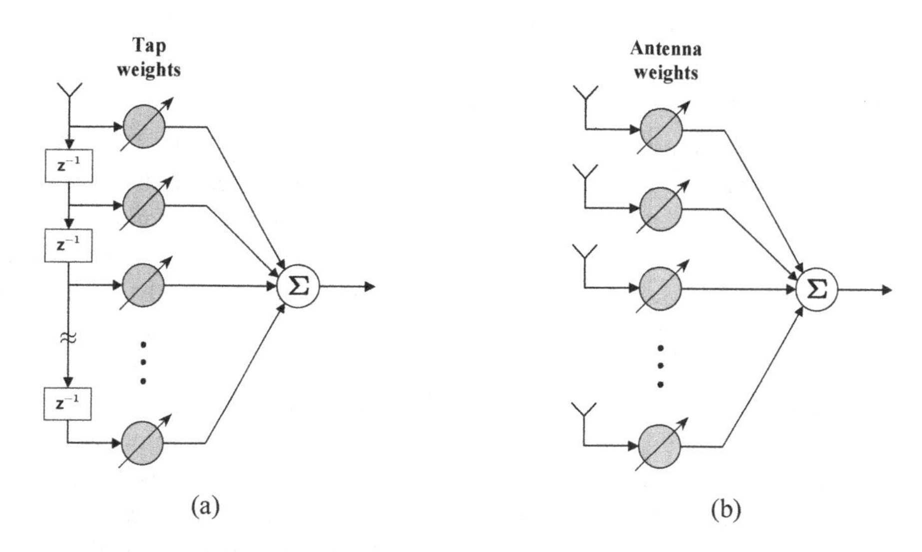

Figure 1.3: Analogy between (a) time-only processing equaliser, and (b) space-only processing beamformer.

Time-only processing generally corresponds to the temporal equalisation of the received signal as illustrated in Figure 1.3a. The equaliser forms a weighted sum of its discrete-time sample outputs corresponding to a single receiving antenna element. The principal behind the equalisation is an attempt to yield a net overall channel response effect that is flat with phase linearity. Time-only processing technique hence offers a very efficient mitigation against channel-induced ISI arising from frequency-selective fading effect. This process of equalising the ISI is in effect gathering the dispersed symbol energy back into its original time intervals so that it will not be lost and be an interference to other symbols. Space-only processing, on the other hand, can be viewed analogously as being the equalisation of the received signal in the spatial domain as shown in Figure 1.3b. The equaliser is realised by implementing more than one antenna element to create a beamformer that takes a weighted sum of its received antenna outputs. The operation of the beamformer basically steers its mainlobe beam towards the direction of a particular user of interest whilst nulling the contributions from the rest of its co-channel users. This therefore makes space-only processing technique an efficient mitigation in suppressing the effects of CCI/MAI in the channel.

In contrast, space-time processing embodies both the ISI and CCI/MAI mitigation superiorities by putting together the two temporal and spatial signal processing techniques. This new spatial-temporal dimension allows interference cancellation to be performed in a way that is not possible with solely the time-only or space-only processing technique. Its classification [46] can be generally categorised into two broad families: decoupled space-time processing and joint space-time processing. The former typically involves a spatial beamformer at the front-end of its configuration followed by a temporal processor. The main idea of decoupled space-time processing [47-52] is to separate CCI/MAI and ISI mitigation into two processing stages. The beamsteering processor first aids in the suppression of the CCI/MAI contributions while preserving all the ISI structure of the desired signal to be exploited later by the temporal equaliser. The latter joint space-time processing [51-58] however exploits the conjoined spatial and temporal signatures of the multipath as a means of signal differentiation. This provides more degrees of freedom than its decoupled counterpart in the suppression of both the ISI and CCI/MAI effects in the channel. As such, its combat against interference is no longer dependent on the individual consideration of the space followed by time differentiation, but on the joint spatial-temporal distinction attributed to each multipath.

In addition to that, space-time processing permits the spatial and temporal consideration of the multipath propagation channel to be incorporated as an inherent source of diversity in the receiver architecture. Diversity is a powerful technique employed extensively in wireless communications network to alleviate the effects of channel fading. Traditionally, this source of diversity can be achieved by transmitting multiple copies of the signal in any of the frequency, time, space or polarisation domains. In other words, replicas of the same information signal can be rendered and transmitted by means of different carrier frequencies, different time slots, different positional spaces, or different states of polarisation. The basic idea of such diversity concept is to create several copies of the same information carrying signal so that each of them will tend to fade independently in the channel; and hence will be less likely for them to suffer simultaneous deep fade at the receiver for any given instant of time. However instead of having the transmitter deliberately generating these multiple copies of the signal, the receiver can employ spatial-temporal processing technique to exploit and derive the source of diversity that is already inherent in the multipath environment. Hence, in summary, space-time processing is capable of not only providing a mitigation of the prejudicial multipath channel effects, but also offering a source of diversity from the multiple signal replicas formed in the propagation channel.

The implementation of space-time processing in wireless communications can provide several significant leverages to enhance the performance of the cellular network system. First of all is the network capacity. As a result of its capability of suppressing the effects of CCI/MAI and ISI in the channel, the number of active users that can be supported in each cell for a given quality of service can thus be significantly increased. This hence gives rise to a gain in its system capacity and brings about an improvement in its spectrum efficiency. Second is the link quality. The quality of the link can be ameliorated by means of combining the diverse multipath contributions inherently present in the otherwise detrimental propagation environment. Finally the third is the range coverage. The high directivity and gain associated with space-time array system allows its corresponding reception range to be extended, hence leading to a larger and better cell coverage. This also entails that lesser transmit power is now required for a given reception distance, consequently resulting in an increase in its transmission efficiency. Besides that, multiple beams can also be employed to dynamically track the transmitters so as to reduce the level of hand-off rate in the cellular network.

1.3 Motivation and Organisation of Book

From the discussions in Sections 1.1 and 1.2, it is therefore conceivable that the integration of space-time array processing and spread spectrum multiple access scheme will bring about a synergetic impact on the overall performance of the wireless communications system. The introduction of the multiple antenna elements, facilitating the application of space-time processing, adds a new dimension in the conception of the spread spectrum estimation/reception process. With the aid of space-time processing, coupled with the exploitation of the DS-CDMA data structure, (1) extra layer of interference cancellation can now be devised, as well as (2) the alleviation of the multipath channel fading effects being instead incorporated as a source of diversity in the receiver design. However the implementation and performance improvements of space-time processing in DS-CDMA scheme are, to a large extent, related to the channel scenario under consideration and the form of diversity that can be derived from the system. Specifically developed space-time vector channel estimation and reception algorithms are thus proposed based on these different environmental context. This hence forms the subject of this written text, which is structured in the outline as follows.

Chapter 2 will first provide an overview of the spatial-temporal array architectural system which is founded on fusing space-time array processing with spread spectrum multiple access scheme. The system model covers from the transmission of the signal, propagation through the multiuser space-time channel, down to its reception at the array front-end. This modelling forms the basic mathematical framework for later formulation in the subsequent chapters.

Chapter 3 introduces the notion of exploiting the polarisation domain of the signal inherent in the environment as a form of diversity and as a means of signal discrimination. This can be achieved by utilising polarisation-sensitive sensor array to capture all the incoming diversely-polarised multipath signals impinging on the sensor elements. More often than not, this diverse polarisation information available in the propagation channel is oftentimes being ignored at the receiver's estimation and reception process. The polarisation of the signal is either assumed to be perfectly aligned with the orientation of the receiving sensor elements, or regarded to be part of the signal fading effect. On the contrary, this work will look at how this polarisation information can be incorporated in the space-time processing of the signal, and the different ways in which the performance of the resulted system can be improved.

Chapter 4 addresses the issue of signal diffusion as a result of the scattering mechanism intrinsical in the propagation environment. The diffuse signal is basically a conglomeration of multiple point sources superimposed together to create the spreading phenomenon as seen in the signal cluster. This hence results in performance degradation if conventional space-time processing techniques based on point sources assumption are applied. In this study, a generalised diffusion framework is attempted in the receiver design in order to handle the occurrences of both point and/or diffuse sources. The algorithm is robust against any incorrect or incomplete erroneous estimates incurring in the estimation process. And due to its underlying structure, the final outcome of the system is also capable of operating in either a co-code (brought about as a consequence of intentional code-reuse scenario or unintentional jamming situation) or a non co-code user environment.

Chapter 5 investigates the derivation of Doppler diversity in an asynchronous Multiple Input and Multiple Output (MIMO) system. Essentially, the Doppler spreading effect, which is normally treated as an impairment factor, constitutes a multiplicity of the signal components dispersed in the frequency domain. By employing the spatial-temporal array architecture at the front-end of the MIMO receiver, instead of the traditional multiple independent-antenna configuration, such channel dispersion can be taken into consideration in the devised space-time processing operation. Additional form of diversity in the Doppler frequency domain can also be derived and be incorporated in the system design. Not only that, the blind near-far resistant MIMO receiver, unlike its conventional MIMO counterparts, does not require the imposition in the need of any power control or any knowledge of the channel.

Chapter 6 looks at some other research works applying the above algorithmic framework in accordance to the context of its environment. These representative examples are included to help in gaining a better understanding and new perspective in the devised space-time vector channel estimation and reception algorithms.

Chapter 7 finally gives an overall concluding view of the current study, and thenceforth identifies potential areas for future research work.

CHAPTER 2

Spatial-Temporal ARray (STAR) Architecture

A general framework of the Spatial-Temporal ARray (STAR) architecture is presented which lays the basic foundation for the subsequent works in this text. Its underlying structure is essentially based on the integration of space-time array processing in spread spectrum multiple access system. By exploiting the spatial-temporal properties of the spread spectrum signal, the mathematical formulation established in the system model forms the basis for the later vector channel estimation and reception applications. To get an overview in the framework of the STAR system model, its primary elemental blocks will be outlined in the next section. Following that, a detailed modelling of each of the blocks will be described in the rest of the remaining sections. This is then followed by a brief summary of the overall system model concluded in the final section.

2.1 Basic Elements of System Model

The fundamental principle that underlies the architecture of the STAR system model relies on two key technologies: sensor array processing and Code Division Multiple Access (CDMA). A functional illustration outlining the basic elements of the STAR communications system is as shown in Figure 2.1.

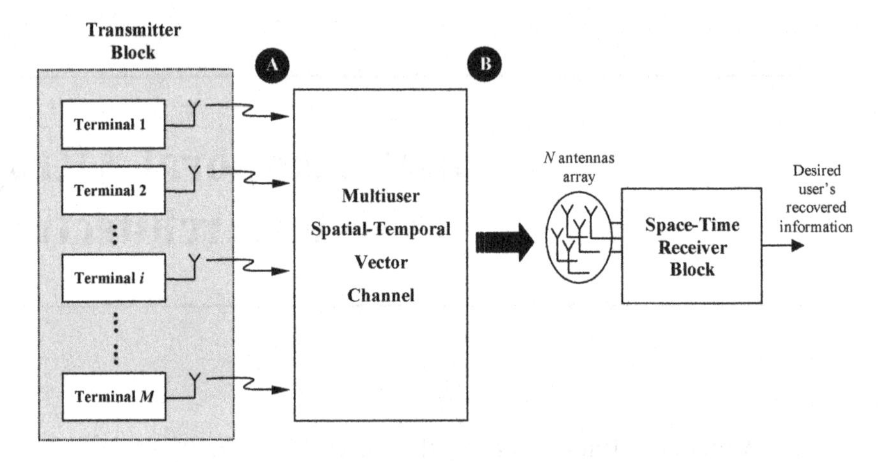

Figure 2.1: Block diagram of a STAR communications system.

The transmitter block, comprises of M transmitting terminals sharing a common channel, constitutes the multiaccess communication setting. Its channel-sharing approach allows multiple transmitting terminals to occupy the same frequency band, in the same space, at the same time by assigning every of its terminals a unique code signature in a random multiple-access CDMA environment. The space-time receiver block will then attempt to recover the desired user's information based on the net received signal vector composed of a superposition of all the transmitters' multipath signals overlapping in the same frequency, space and time domains. To evoke the receiver's space-time array processing capability, a sensor array consisting of N antenna elements is introduced at the front-end of the receiver. The link between the transmitter and receiver blocks (between point A and B) is then provided for by the physical transmission medium, referred herein as the multiuser spatial-temporal vector channel. Of particular importance in the study of the STAR architecture is the characteristic of this channel as it generally affects the design of the primary building blocks of the space-time array receiver. For a better and more in-depth understanding, the mathematical formulation behind each of the basic elements of the STAR communications system will be delineated in the next few sections.

2.2 Spread Spectrum Transmission

Since the multiple-access environment as depicted in Figure 2.1 encompasses the case of a common transmitting terminal for all the users, for ease in explanation, let's just consider the working operation of the ith user's terminal in an asynchronous DS-CDMA system. Note that the input baseband signal for each of the terminals may be from a voice/ video/data source. Its digital representation is compressed by means of a source encoder, followed by a channel encoder which introduces some redundancy in the message sequence to combat the effect of noise and interference encountered in the channel. However to portray the actual performance of the proposed works, such added redundancy to improve the reliability and fidelity of the signal will not be considered in this study. Also, the asynchronous operation of the transmitting terminal, unlike its synchronous counterpart, does not have the advantage of having symbol synchronism via a common clock. To see this, a pictorial illustration depicting the basic operation of the ith user's direct sequence spread spectrum transmitter is as shown in Figure 2.2.

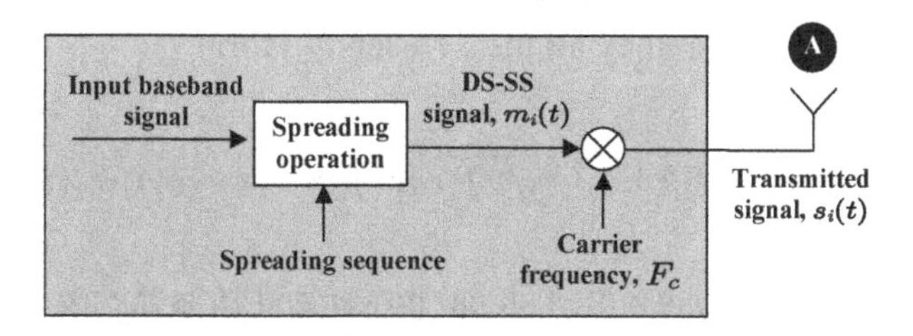

Figure 2.2: A direct sequence spread spectrum transmitter.

As seen from the figure, the input narrowband baseband signal is first multiplied by a wideband spreading waveform, which is composed of a sequence of chips, to obtain the following Direct Sequence Spread Spectrum (DS-SS) message signal

$$m_i(t) = \sum_{n=-\infty}^{+\infty} \mathrm{a}_i[n] c_{PN,i}(t - nT_{cs}), \quad nT_{cs} \leq t < (n+1)T_{cs} \tag{2.1}$$

where $\{\mathrm{a}_i[n], \forall n \in \mathcal{Z}\}$ is the ith user's data symbol, and T_{cs} is the channel symbol period. Note that the sequence of chips not only serves as a spreading function, but also as an identification code among the different users. Its pseudo-noise (PN) spreading waveform associated with the ith user, $c_{PN,i}(t)$, is modelled as

$$c_{PN,i}(t) = \sum_{m=0}^{\mathcal{N}_c - 1} \alpha_i[m] c(t - mT_c), \quad mT_c \leq t < (m+1)T_c \tag{2.2}$$

where $\{\alpha_i[m] \in \{-1, +1\}, m = 0, 1, \ldots, \mathcal{N}_c - 1\}$ corresponds to the ith user's PN-code sequence, and $c(t)$ denotes the unit amplitude chip pulse-shaping waveform of duration T_c which is zero outside the range $0 \leq t < T_c$. Its associated spreading factor, or processing gain, is given by $\mathcal{N}_c = T_{cs}/T_c$ which is the periodical length of the ith user's spreading code sequence vector $\underline{\alpha}_i = [\alpha_i[0], \ \alpha_i[1], \ldots, \ \alpha_i[\mathcal{N}_c - 1]]^T$.

Prior to transmission at point A, the spread spectrum message signal $m_i(t)$ is upconverted and translated to a common carrier frequency F_c, shared by all the M users, to form

$$s_i(t) = \sqrt{2P_i} \exp(j(2\pi F_c t + \vartheta_i)) m_i(t) \tag{2.3}$$

where P_i is the transmitted signal power and ϑ_i is the associated random carrier phase term uniformly distributed between $[0, \ 2\pi)$.

2.3 Multiuser Space-Time Vector Channel

To understand the space-time multipath vector channel, it is necessary to have a preliminary overview of the array manifold vector due to the antenna array attached at the front-end of the receiver. The receiver's manifold vector forms part of the components in the modelling of the vector channel, alongside other intrinsic channel parameters in the propagation environment. With that overview, a stepwise formulation in the modelling of the multiuser spatial-temporal vector channel can thence be constructed. Next, the transmitted signal at point A is then mapped, using the devised vector channel model, to realise its corresponding continuous-time signal vector appearing at the receiver's frontal array elements at point B.

2.3.1 Sensor Array Manifold Vector

The spatial array manifold vector, sometimes referred to as the array response vector, is a function consisting of the sensor array geometry, the signal carrier frequency and the path angle-of-arrival. Consider a receiver having an antenna array of N omnidirectional sensors. Its array geometrical location is given by $[\underline{r}_x, \underline{r}_y, \underline{r}_z] \in \mathcal{R}^{N \times 3}$ with each row denoting the Cartesian coordinates of each of the sensor elements. Suppose an incoming multipath signal of wavelength λ arrives at the receiver from an angular direction of (θ, ϕ), with $\theta \in [0, 2\pi)$ and $\phi \in [-\pi/2, \pi/2]$ denoting its azimuth and elevation angle respectively. Its associated sensor array manifold vector pertaining to that signal path can be expressed explicitly as

$$\underline{S}(\theta, \phi) \quad = \quad \exp\big(-j\,[\,\underline{r}_x, \underline{r}_y, \underline{r}_z\,]\,\underline{k}(\theta, \phi)\big) \tag{2.4}$$

where $\underline{k}(\theta, \phi) = (2\pi/\lambda).\underline{u}$ is the wavenumber vector with the unit vector $\underline{u} \triangleq \underline{u}(\theta, \phi) = [\cos(\theta).\cos(\phi),\ \sin(\theta).\cos(\phi),\ \sin(\phi)]^T$.

Without any loss of generality, this study will adopt the assumption of coplanar signals impinging on the array with elevation angle $\phi = 0$. The locus traced by all its manifold vectors, which is now a function of the single parameter θ, constitutes the array manifold. This 1-D manifold curve, embedded in an N-dimensional complex space, forms a vector continuum [59] for $\forall\theta$, as pictorially represented in Figure 2.3.

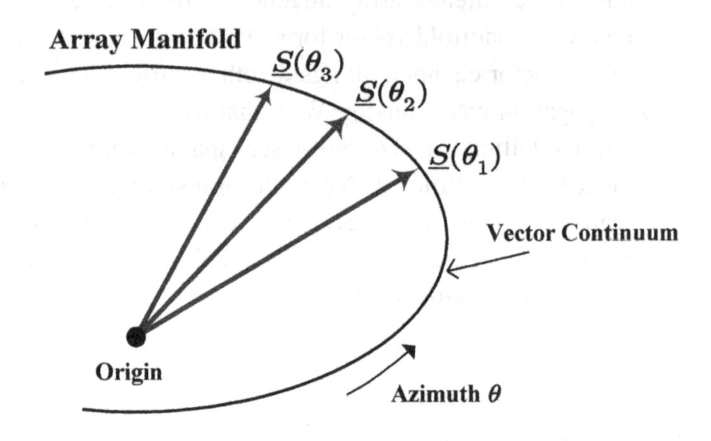

Figure 2.3: Array manifold of omnidirectional antenna array (vector continuum).

2.3.2 Vector Channel Mathematical Model

The signal transmission of each user, propagating through the channel, is assumed to have experienced multipath fading and dispersion in the medium. A vector channel model is used to characterise this phenomenon with each of its multipath components, in the propagation channel, being associated with a particular Direction-of-Arrival (DOA) and Time-of-Arrival (TOA). Suppose the transmitted signal due to the ith user arrives at the receiver via K_i multipaths, its corresponding Vector Channel Impulse Response between the transmitter and receiver, that is from point A to point B, can be written as

$$\underline{\mathcal{I}}_i(t) \quad = \quad \sum_{j=1}^{K_i} \beta'_{ij}\, \underline{S}(\theta_{ij})\, \delta(t - \tau_{ij}) \qquad (2.5)$$

where $\underline{S}(\theta_{ij})$ is the array manifold vector pointing at azimuth θ_{ij}

direction, β'_{ij} is the complex fading coefficient encompassing the path's amplitude and random phase shift, and τ_{ij} is the associated signal path delay, for the jth path of the ith user.

2.3.3 Continuous-Time Signal Model

The Vector Channel Impulse Response is useful in mapping the transmitted signal from a specific user at point A to the received arrayed-signal at point B. It stipulates the conjoined spatial and temporal signature for each of the multipath components which is the basis to be exploited later in the receiver's space-time estimation/reception process.

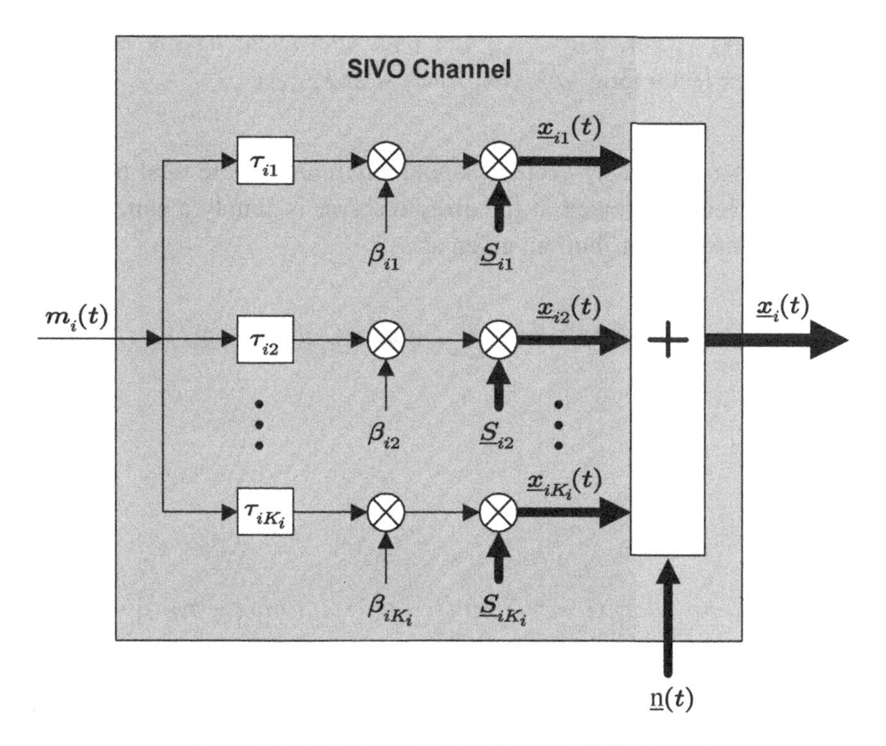

Figure 2.4: Scalar-Input Vector-Output (SIVO) channel.

To see the mapping operation provided by the Vector Channel Impulse Response, a diagrammatic interpretation, after removal of the carrier, is portrayed in a Scalar-Input Vector-Output (SIVO) channel as shown in Figure 2.4. Hence given a transmitted signal from the ith user inputted into the channel, the mapping operation will generate accordingly a set of multipath arrayed-signal at the front-end of the receiver. Its resulted net baseband vector representation of the received signal, in the presence of additive isotropic white Gaussian noise with double sided power spectral density $N_0/2$, can thus be expressed as

$$\underline{x}_i(t) = \sum_{j=1}^{K_i} \beta_{ij}\, \underline{S}(\theta_{ij})\, m_i(t-\tau_{ij}) + \underline{n}(t) \tag{2.6}$$

where $\underline{n}(t) \in \mathcal{C}^{N\times 1}$ is the complex additive white Gaussian noise (AWGN) vector, and β_{ij} is the path coefficient having absorbed the complex factor term $\sqrt{2P_i}\exp(\,j(\vartheta_i - 2\pi F_c \tau_{ij}))$.

Now for a multiuser vector channel environment, the total net baseband signal vector received at the array receiver is simply a superposition of all the users' contribution, given as

$$\underline{x}(t) = \sum_{i=1}^{M} \mathbb{S}_i \operatorname{diag}(\underline{\beta}_i)\, \underline{m}_i(t) + \underline{n}(t) \tag{2.7}$$

where

$$\begin{aligned}
\mathbb{S}_i &= \left[\underline{S}_{i1},\ \underline{S}_{i2},\ \ldots,\ \underline{S}_{iK_i}\right] && \in \mathcal{C}^{N\times K_i}\\
\underline{\beta}_i &= \left[\beta_{i1},\ \beta_{i2},\ \ldots,\ \beta_{iK_i}\right]^T && \in \mathcal{C}^{K_i\times 1}\\
\underline{m}_i(t) &= \left[m_i(t-\tau_{i1}),\ m_i(t-\tau_{i2}),\ \ldots,\ m_i(t-\tau_{iK_i})\right]^T && \in \mathcal{C}^{K_i\times 1}
\end{aligned}$$

with the array manifold vector defined as $\underline{S}_{ij} \triangleq \underline{S}(\theta_{ij})$. For a clearer understanding, the mathematical formulation associated with the above expression is described in a Vector-Input Vector-Output (VIVO) channel model as illustrated in Figure 2.5.

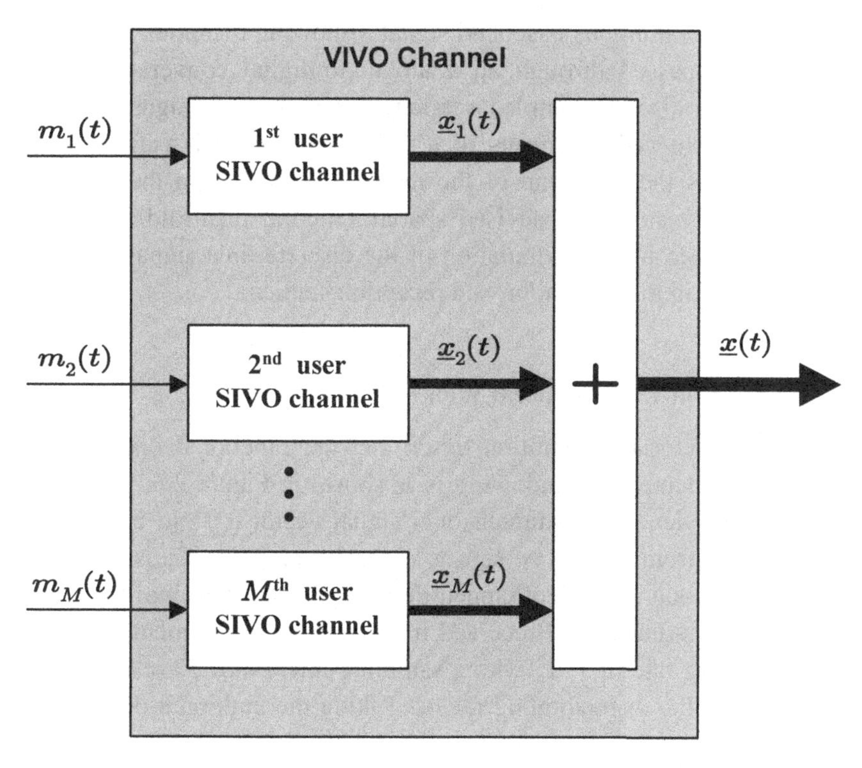

Figure 2.5: Vector-Input Vector-Output (VIVO) channel.

Alternatively, the continuous-time signal vector expression delineated in Equation (2.7) can be rewritten, in a more compact manner, as

$$\underline{x}(t) \quad = \quad \mathbb{S} \,.\, \text{diag}\big(\underline{\beta}\big) \,.\, \underline{m}(t) + \underline{\text{n}}(t) \tag{2.8}$$

$$\text{where} \begin{cases} \mathbb{S} & = \quad \big[\mathbb{S}_1, \quad \mathbb{S}_2, \quad \ldots, \quad \mathbb{S}_M\big] \\ \underline{\beta} & = \quad \big[\underline{\beta}_1^T, \quad \underline{\beta}_2^T, \quad \ldots, \quad \underline{\beta}_M^T\big]^T \\ \underline{m}(t) & = \quad \big[\underline{m}_1^T(t), \quad \underline{m}_2^T(t), \quad \ldots, \quad \underline{m}_M^T(t)\big]^T \end{cases}$$

2.4 Space-Time Array Reception

The continuous baseband signal, following reception at the antenna array, is passed through an analogue-to-digital conversion (ADC) to obtain the discrete sampled version of the baseband signal vector. This can be achieved by means of a front-end temporal windowing which effectuates the inclusion of the temporal term within the spatial array manifold vector. The devised spatial-temporal manifold vector is then incorporated in the formulation of the discrete-time signal vector, to be used later in the estimation and reception studies.

2.4.1 Front-End Temporal Windowing

A pictorial illustration describing the functional operation of the front-end temporal windowing is as shown in Figure 2.6. Upon removal of the carrier, the N-dimensional signal vector $\underline{x}(t)$ in Equation (2.8), obtained from the array sensor elements at point C, is sampled and passed through a bank of tapped-delay lines (TDL) each of length L. The continuous-time signal received from each sensor element is sampled at a sampling rate of $1/T_s$ with a sampling interval of $T_s = T_c/q$ in which $q \in \mathcal{N}$ is the oversampling factor. Taking the multipath delay spread to be within the range $[0, T_{cs})$, the path's delay τ_{ij} can therefore be discretised into its integer delay $l_{ij}T_s$ and fractional delay $\varrho_{ij}T_s$, where $l_{ij} \in \{0, 1, \ldots, T_{cs}/T_s - 1\}$ and $\varrho_{ij} \in [0, 1)$. The acquired samples are then fed into the TDL bank to obtain a temporal windowing of length $L = 2q\mathcal{N}_c$. Upon concatenating its output at point D, a NL-dimensional discretised signal vector is thus formed and read for every T_{cs}, with the nth observation interval given as

$$\underline{x}[n] \quad = \quad \left[\underline{x}_1[n]^T, \underline{x}_2[n]^T, \ldots, \underline{x}_N[n]^T\right]^T \tag{2.9}$$

where $\underline{x}_k[n]$ is the L-dimensional output frame from the kth TDL, i.e.

$$\underline{x}_k[n] = \left[x_k[nT_{cs}], x_k[nT_{cs} - T_s], \ldots, x_k[nT_{cs} - (L-1)T_s]\right]^T \tag{2.10}$$

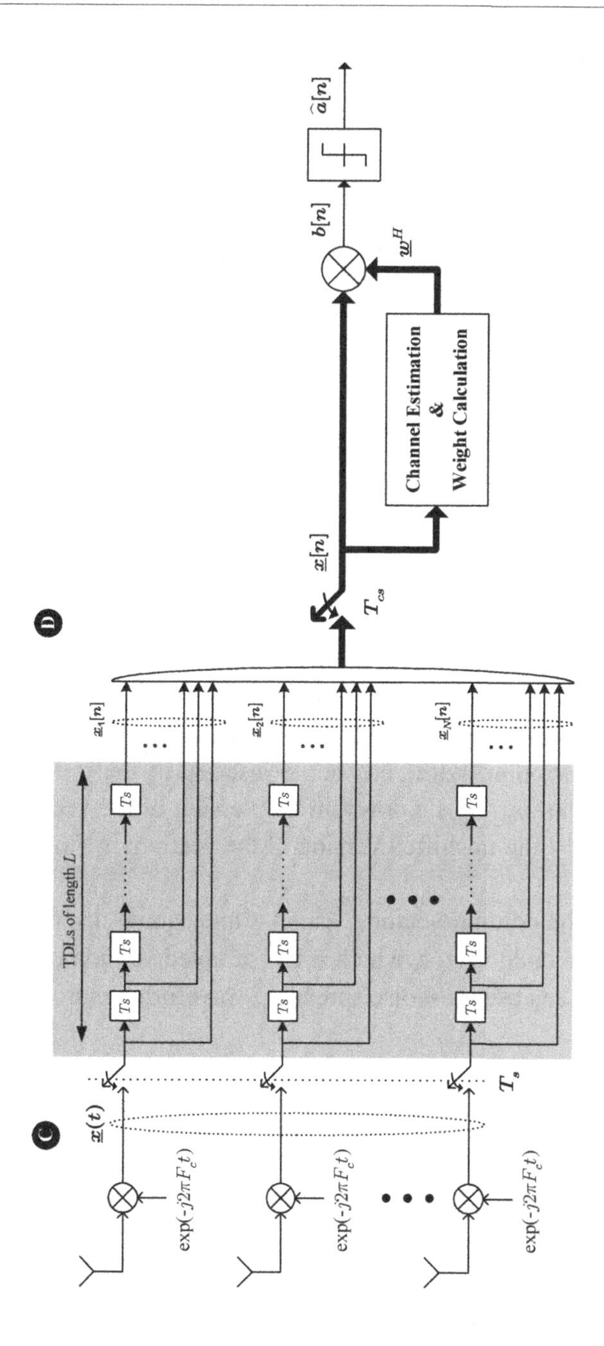

Figure 2.6: Front-end temporal windowing.

2.4.2 Spatial-Temporal Array Manifold Vector

In order to model the discretisation of the received complex signal vector, it is essential to first define the Spatial-Temporal ARray (STAR) manifold vector. The manifold vector is dependent on two terms: (1) the spatial array manifold vector, and (2) the temporally-delayed code vector, for every j^{th} path of the ith user. The latter is formulated as

$$\underline{c}_{ij} = \mathbb{J}^{l_{ij}} \underline{c}_i \tag{2.11}$$

where $l_{ij} = \lceil \tau_{ij}/T_s \rceil$ is the discretised path delay and $\mathbb{J}$ (or $\mathbb{J}^T$) is a $L \times L$ time down-shift (or up-shift) operator matrix given as

$$\mathbb{J} = \begin{bmatrix} 0 & 0 & \cdots & 0 & 0 \\ 1 & 0 & \cdots & 0 & 0 \\ 0 & 1 & \cdots & 0 & 0 \\ \vdots & \vdots & \ddots & \vdots & \vdots \\ 0 & 0 & \cdots & 1 & 0 \end{bmatrix} = \begin{bmatrix} \underline{0}_{L-1}^T & 0 \\ \mathbb{I}_{L-1} & \underline{0}_{L-1} \end{bmatrix} \tag{2.12}$$

Remark: Notice that every time the matrix $\mathbb{J}$ (or $\mathbb{J}^T$) operates on a column vector, the contents of the vector are downshifted (or upshifted) by one position, with zero being inserted at the top (or bottom) of the vector; that is, $\mathbb{J}^l \underline{c}$ is a downshifted version of the vector $\underline{c}$ by l elements, and $(\mathbb{J}^T)^l$ is an upshifted version of the vector $\underline{c}$ by l elements.

Lastly, the column vector $\underline{c}_i$ is a L-dimensional discretised zero-delayed reference code vector, which is the sampled version of a single period of the ith user's pseudo-noise spreading waveform, expressed as

$$\underline{c}_i = \begin{bmatrix} c_{PN,i}(0) \\ c_{PN,i}(T_s) \\ \vdots \\ c_{PN,i}((q\mathcal{N}_c - 1)T_s) \\ 0 \\ 0 \\ \vdots \\ 0 \end{bmatrix} = \sum_{p=0}^{\mathcal{N}_c - 1} \alpha_i[p] \,.\, \mathbb{J}^{pq} \underline{c} \tag{2.13}$$

in which $\underline{c}$ is the oversampled chip-level pulse shaping function $c(t)$ padded with zeros at the end, i.e.

$$\underline{c} = \left[c(0),\ c(T_s),\ \cdots,\ c((q-1)T_s),\ \underline{0}^T_{(L-q)}\right]^T \tag{2.14}$$

It is worthwhile to note that in the case of a chip-rate sampler of $q = 1$, as employed in this study, the temporal windowing length becomes $L = 2\mathcal{N}_c$ and its reference code vector $\underline{\mathfrak{c}}_i$ is reduced to

$$\underline{\mathfrak{c}}_i = \Big[\underbrace{\left[\alpha_i[0],\ \alpha_i[1],\ \cdots,\ \alpha_i[\mathcal{N}_c - 1]\right]}_{i^{\text{th}}\text{ user's PN-code}},\ \underline{0}^T_{\mathcal{N}_c}\Big]^T \tag{2.15}$$

which is basically a makeup of the ith user's PN-code sequence padded with $\mathcal{N}_c$ zeros at the end.

Now let's look at the incorporation of the above temporally-delayed code vector together with the sensor array manifold vector. Recall that the array manifold vector is previously employed in the spatial mapping of the signal path across the antenna elements of the array, as depicted in Figure 2.4. The Spatial-Temporal ARray (STAR) manifold vector, on the other hand, is purposed to map each of the signal paths in both the spatial and temporal domains. To illustrate this, consider the signal path due to the jth path of the ith user. Its associated STAR manifold vector can be found by combining the spatial array manifold vector and the discretised temporally-delayed code vector as follows

$$\begin{aligned} \underline{\mathfrak{h}}_{ij} &= \underline{S}_{ij} \otimes \underline{\mathfrak{c}}_{ij} \\ &\triangleq \underline{S}_{ij} \otimes \mathbb{J}^{l_{ij}} \underline{\mathfrak{c}}_i \end{aligned} \tag{2.16}$$

with $\otimes$ denoting the Kronecker product. Such conjoint STAR manifold vector exploits jointly the spatial-temporal signature of each signal path which is the key in the conceptualisation of the parametric space-time channel estimation process. The locus inscribed by this STAR manifold vector is unique for every different user's spreading code sequence. Hence the STAR manifold locus of one user will be distinct from that of another user. This thus provides a basis of distinguishing the multipath signals due to each individual user operating in a multiuser environment.

2.4.3 Discrete-Time Signal Model

Having got the STAR manifold vector, which characterises both the spatial and temporal domains of the multipath signals, the discretised space-time channel response vector for the nth symbol of the ith user can be deduced from Equation (2.5) to obtain

$$\begin{aligned} \underline{\mathcal{I}}_i[n] &= \sum_{j=1}^{K_i} \beta_{ij} \underline{\mathfrak{h}}_{ij} \\ &= \left[\underline{\mathfrak{h}}_{i1}, \underline{\mathfrak{h}}_{i2}, \ldots, \underline{\mathfrak{h}}_{iK_i}\right] \cdot \begin{bmatrix} \beta_{i1} \\ \beta_{i2} \\ \vdots \\ \beta_{iK_i} \end{bmatrix} \end{aligned} \tag{2.17}$$

As with the mathematical model for the Vector Channel Impulse Response, the above discretised channel response vector is simply a summation of the ith user's multipath components. However the above expression only pertains to the contributions associated with the current symbol $a_i[n]$. As a result of the multipath delay spread, the content in each of the TDLs contains contributions from not only the current but also the previous and next symbols. To see this, take for instance the three multipath signals, due to the ith user, with each arriving at a time delay of l_{i1}, l_{i2} and $l_{i3} \in [0, T_{cs}/T_s)$ respectively. For a delay of $l_{i1} < l_{i3} < l_{i2}$ as depicted in Figure 2.7, the misalignment in the symbol epochs within the TDL will thenceforth constitute contributions from the previous and next symbols, which consequently leads to the Inter-Symbol Interference (ISI) effect.

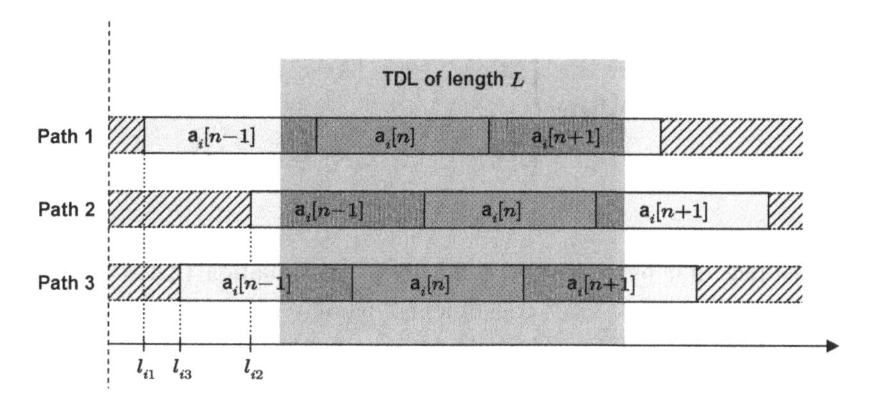

Figure 2.7: Inter-Symbol Interference (ISI) resulted from previous and next symbols.

Now for the sake of clarity, let's just consider for the moment, the discrete-time signal model associated with only the contribution from the current symbol. By employing the STAR manifold vector as shown in Equation (2.16), the discretised signal model for all the M users as a result of the current symbol's contribution may therefore be derived as

$$\underline{x}[n] \quad = \quad \sum_{i=1}^{M} \mathbb{H}_i \, \underline{\beta}_i \, \mathrm{a}_i[n] + \underline{\mathrm{n}}[n] \tag{2.18}$$

where the matrix $\mathbb{H}_i = \left[\underline{\mathfrak{h}}_{i1}, \underline{\mathfrak{h}}_{i2}, \ldots, \underline{\mathfrak{h}}_{iK_i}\right]$ has columns containing the STAR manifold vectors of the K_i multipaths due to the ith user, and $\underline{\mathrm{n}}[n] \in \mathcal{C}^{NL\times 1}$ is the sampled noise vector.

Next, by taking into account all the contributions from the previous, current and next symbols, the spatial-temporal discretised signal vector $\underline{x}[n]$ will thus become

$$\underline{x}[n] \quad = \quad \sum_{i=1}^{M} \left[\mathbb{H}_i^{\text{prev}} \underline{\beta}_i, \mathbb{H}_i \underline{\beta}_i, \mathbb{H}_i^{\text{next}} \underline{\beta}_i\right] \begin{bmatrix} \mathrm{a}_i[n-1] \\ \mathrm{a}_i[n] \\ \mathrm{a}_i[n+1] \end{bmatrix} + \underline{\mathrm{n}}[n] \tag{2.19}$$

$$\text{where} \begin{cases} \mathbb{H}_i^{\text{prev}} = \left(\mathbb{I}_N \otimes \left(\mathbb{J}^T\right)^{L/2}\right)\mathbb{H}_i \\ \mathbb{H}_i^{\text{next}} = \left(\mathbb{I}_N \otimes \mathbb{J}^{L/2}\right)\mathbb{H}_i \end{cases}$$

Notice that by rearranging the terms in Equation (2.19), $\underline{x}[n]$ can be decoupled into four constituents, namely the desired, Inter-Symbol Interference (ISI), Multiple-Access Interference (MAI) and noise components. With the $\mathcal{D}^{\text{th}}$ user taken as the desired user of interest, Equation (2.19) can hence be rewritten to obtain

$$\underline{x}[n] = \mathbb{H}_{\mathcal{D}}\underline{\beta}_{\mathcal{D}}\mathrm{a}_{\mathcal{D}}[n] + \underline{I}_{\text{ISI}}[n] + \underline{I}_{\text{MAI}}[n] + \underline{\mathrm{n}}[n] \quad (2.20)$$

$$\text{where} \begin{cases} \mathbb{H}_{\mathcal{D}}\underline{\beta}_{\mathcal{D}}\mathrm{a}_{\mathcal{D}}[n] \text{ is the desired signal component} \\ \underline{I}_{\text{ISI}}[n] = \left[\mathbb{H}_{\mathcal{D}}^{\text{prev}}\underline{\beta}_{\mathcal{D}}, \mathbb{H}_{\mathcal{D}}^{\text{next}}\underline{\beta}_{\mathcal{D}}\right]\begin{bmatrix} \mathrm{a}_{\mathcal{D}}[n-1] \\ \mathrm{a}_{\mathcal{D}}[n+1] \end{bmatrix} \\ \underline{I}_{\text{MAI}}[n] = \sum_{\substack{i=1 \\ i\neq\mathcal{D}}}^{M} \left[\mathbb{H}_i^{\text{prev}}\underline{\beta}_i, \mathbb{H}_i\underline{\beta}_i, \mathbb{H}_i^{\text{next}}\underline{\beta}_i\right]\begin{bmatrix} \mathrm{a}_i[n-1] \\ \mathrm{a}_i[n] \\ \mathrm{a}_i[n+1] \end{bmatrix} \end{cases}$$

The motivation behind the space-time array reception is therefore to devise a methodology of extracting the desired signal component so as to facilitate a recovery of its original transmitted data symbol, as illustrated in Figure 2.6 Specifically-designed space-time vector channel estimation and reception techniques, based on the net received discretised signal vector, are thus developed under this study. Each of these will be further elaborated in the subsequent chapters of the text.

2.5 Summary

The STAR communications system is based on an integration of two key technologies, namely sensor array processing and DS-CDMA multiple-access scheme. Its architecture is made up of three basic elements: (1) transmitter block, (2) multiuser spatial-temporal vector channel, and (3) space-time receiver block. In particular, the mathematical model of the space-time vector channel, established within the STAR communications framework, plays a significant role in mapping the transmitted signal at the transmitters' end to the received signal at the receiver's end. It aids in the characterisation of the multipath propagation environment by incorporating the conjoint STAR manifold vector which includes both the spatial and temporal dimensions of the multipath signal. This space-time signature inherent in the channel can thus be exploited within the vector channel estimation and reception process, which will be described in the studies that follows.

CHAPTER 3

Polarisation-Space-Time Estimation and Reception

In this work the use of a crossed-dipole array is proposed in joint space-time channel estimation for asynchronous multipath DS-CDMA systems. The polarisation diversity offered by such an array, unlike linearly polarised arrays, is able to detect and estimate any arbitrary polarised signal paths. By utilising the polarisation information inherent in the received signal, the accuracy and resolution of the estimation is significantly improved, and its signal detection capability is enhanced. To alleviate the need for a multidimensional search into the polarisation space, a computationally efficient polarisation-angle-delay estimation (PADE) algorithm is proposed which provides a joint space-time estimate of the desired user in an asynchronous multipath environment. Its polarisation parameters, if required, can also be obtained by solving a set of analytical equations. The proposed algorithm which acts as the front-end estimator for the Polarisation-Spatial-Temporal ARray (Polar-STAR) receiver is supported by representative examples and computer simulation studies.

3.1 Introductory Background

In wireless communications, the signal emitted by a mobile terminal normally suffers multiple reflection, refraction, diffraction and scattering along the transmission path, hence creating several replicas with different arriving angle, path delay, polarisation and fading. However, among these four channel parameters, the polarisation factor of the signal, which describes the orientation of its electric field, often receives disproportionately little attention in traditional model of array processing. Most array processing techniques assume the employment of polarisation-insensitive sensors, which therefore presume that the polarisation of the received signal is perfectly aligned with respect to the orientation of the sensors, thus obviating any polarisation mismatches. But in typical mobile environments, the received signal rarely takes on its transmitted polarisation due to the depolarisation mechanism [41] intrinsical in the propagation channel (especially in an urban environment). This is further aggravated by the frequent random angular orientation of most portable handheld devices. Such diversity in signal's polarisation, which is normally treated as part of signal fading (i.e. polarisation fading), can be exploited to provide an extra degree of signal discrimination, and as such improve the receiver's detection/ estimation capabilities. Note that the polarisation state of the signal can be either completely or partially polarised, but as with many studies, the main focus of this work will assume complete polarisation. Investigation on partially polarised scenario can be found in [60,61].

Polarisation diversity has been studied in a number of direction finding algorithms to ameliorate its angle-of-arrival estimation [62-66]. This is achieved by means of diversely polarised arrays which are sensitive to the polarisation of the received signal. Ferrara and Parks [62] have shown that by employing an array of diversely polarised sensors, the angle-of-arrival estimation is significantly improved as multiple signals can now be resolved on the basis of polarisations in addition to their arriving angles. Schmidt [63], on the other hand, demonstrated the ability of distinguishing two highly correlated signals by incorporating the polarisations of the received signal. Since then, various diversely polarised arrays have been proposed in a correlated signal environment. For instance in [64], a diversely polarised array consisting of circularly-polarised sensors is used with the Cramer-Rao bound to evaluate the angle estimation accuracy of correlated signals. In [65], a linear array of crossed-dipoles, which measures its horizontal and vertical responses separately, is used with the ESPRIT algorithm to estimate the angle and polarisation of coherent signals. The performance of the angle and polarisation estimation is then further improved in [66] by using a co-centered orthogonal loop and dipole (COLD) array. Other diversely polarised arrays include, for instance, the use of dipole triad and/or loop triad for multisource angle and polarisation estimation [67]. Note that a dipole triad (or loop triad) consists of a collocation of three identical and orthogonally-conjoined electrically-induced dipoles (or magnetically-induced loops). In [68], the concept of electromagnetic vector-sensor is employed for self-initiating MUSIC-based direction finding and polarisation estimation algorithm in the spatio-polarisational beamspace. An electromagnetic vector-sensor is comprised of six spatially-collocated nonisotropic antennas where the signal's three electric-field components and three magnetic-field components are each measured separately. These electromagnetic vector-sensors were also employed in [69] for ESPRIT-based blind beamforming or geolocation of wideband fast frequency-hopping signals. However, little work [70] has been done to include polarisation in the area of joint space-time channel estimation [71-73], especially in a correlated signal environment. Hence in this study, a novel joint direction-of-arrival and time-of-arrival estimation approach is proposed for asynchronous DS-CDMA systems, in conjunction with analytical expressions providing the estimate of the polarisation parameters [74-76].

In Section 3.2, the well-known spatial array manifold vector commonly-used for polarisation-insensitive array will first be extended to include the polarisation aspect to form its corresponding polarisation-sensitive manifold vector. Next the temporal dimension is incorporated to construct the Polarisation-Spatial-Temporal ARray (Polar-STAR) manifold vector for the asynchronous DS-CDMA systems. To realistically model the multipath channel using the polarisation-sensitive sensors, instead of the traditional isotropic sensors, a generalised signal model is formulated as detailed in Section 3.3. Based on the formulation, a subspace-based polarisation-angle-delay estimation (PADE) algorithm is then proposed in Section 3.4, together with a novel temporal smoothing technique which is devised to restore the desired subspace dimensionality. Following that, Section 3.5 provides several simulation studies which depict the performance of the proposed algorithm. The work is finally concluded in Section 3.6.

3.2 Polarised Array Manifold Vector

As described in Section 2.3.1, the sensor array manifold vector for an antenna array consisting of N omnidirectional sensors can be expressed as shown in Equation (2.4). The construction of the expression is based on the assumption that all the sensor elements are polarisation-insensitive. In other words, they do not take into account the polarisation of the incoming signals arriving at the array. To consider for the signal's polarisations, the expression needs to be extended to the case of diversely-polarised array by first specifying the polarisation of the signal path. Following that, the formulation of the sensor's response with regards to the polarised signal can then be developed.

3.2.1 Polarisation of Electromagnetic Wave

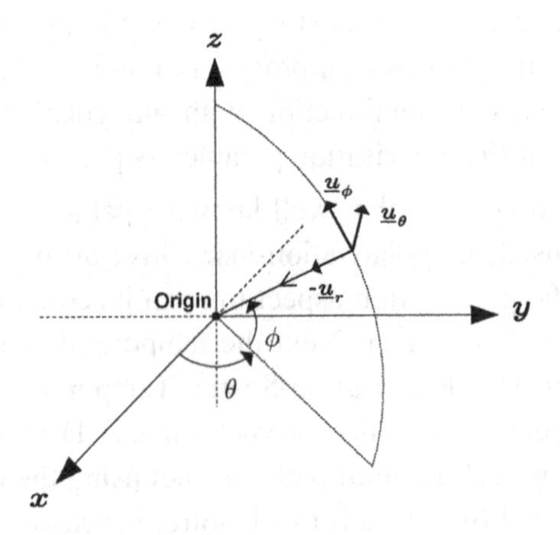

Figure 3.1: An incoming signal towards a diversely-polarised sensor.

Consider a diversely-polarised sensor, as depicted in Figure 3.1, located at the origin of a spherical coordinate system, with the unit vectors $\underline{u}_\phi$, $\underline{u}_\theta$ and $\underline{u}_r$ forming the right-hand coordinate arrangement for an incoming signal. For a given transverse electromagnetic (TEM) wave impinging on the sensor, the polarisation ellipse described by its electric field, as viewed from the coordinate origin, can be illustrated as shown in Figure 3.2.

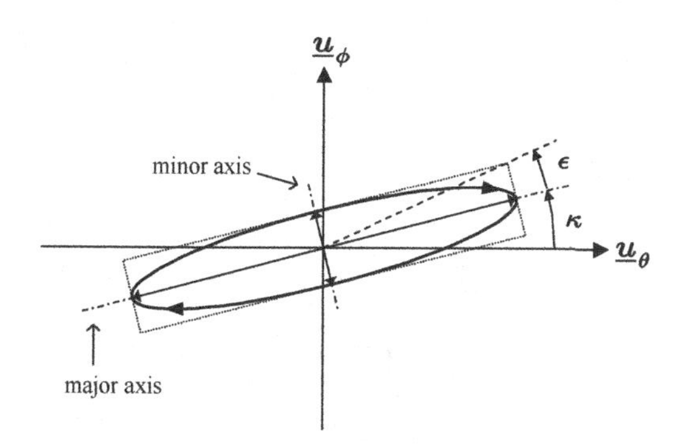

Figure 3.2: Polarisation ellipse described by the electric field.

The electric field of the TEM wave is composed of two transverse components, which are its vertical field element E_ϕ and its horizontal field element E_θ. The polarisation ellipse described by this field can be characterised in terms of the ellipse's orientation and shape. The former is defined by the orientation angle of the ellipse $0 \leq \kappa < \pi$, measured from the horizontal axis $\underline{u}_\theta$ to the ellipse major axis. The latter, on the other hand, is specified by the ellipse axial ratio given as

$$\varpi = \frac{\text{length of minor axis}}{\text{length of major axis}} \tag{3.1}$$

By relating the axial ratio to the ellipticity angle ϵ of the ellipse, the ellipse's shape can be mathematically delineated as follows

$$\epsilon = \tan^{-1}\varpi \tag{3.2}$$

where $-\pi/4 \leq \epsilon \leq \pi/4$ with its sign indicating the sense of rotation.

The above parameters, orientation angle κ and ellipticity angle ϵ, completely characterise the polarisation state of the TEM wave. By translating these parameters onto a Poincaré sphere, they represent the latitude 2κ and longitude 2ϵ of a particular point on the sphere, which consequently defines the signal's state of polarisation as exemplified in Figure 3.3. Alternatively, this one-to-one correspondence can be specified with an alternative set of parameters Gamma γ and Eta η by applying the formulas of spherical trigonometry on the Poincaré sphere.

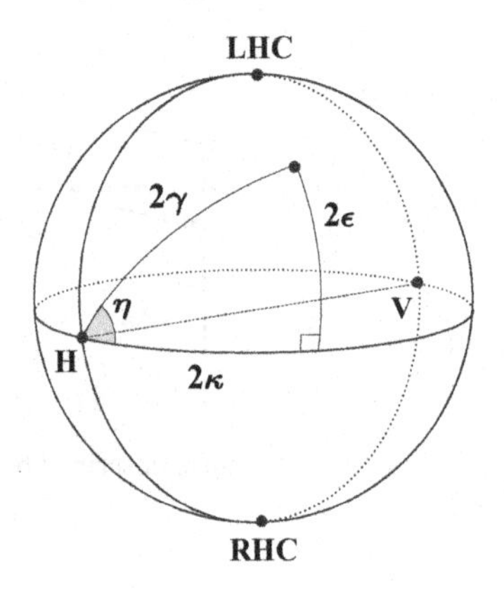

Figure 3.3: Poincaré sphere concept.

The relationship [77] among these four angular parameters γ, η, κ and ε can be described as

$$\cos(2\gamma) = \cos(2\epsilon)\,.\cos(2\kappa) \tag{3.3}$$

$$\tan(\eta) = \tan(2\epsilon)\,.\csc(2\kappa) \tag{3.4}$$

where $0 \leq \gamma \leq \pi/2$ and $-\pi \leq \eta < \pi$. Its inverse relations are given by

$$\tan(2\kappa) = \tan(2\gamma)\,.\cos(\eta) \tag{3.5}$$

$$\sin(2\epsilon) = \sin(2\gamma)\,.\sin(\eta) \tag{3.6}$$

Now for an incoming TEM wave of an arbitrary polarisation state propagating towards the diversely-polarised sensor, its electric field of amplitude $\|\underline{E}\|$ can be described in terms of its transverse components, defined as

$$\underline{E} = E_{\phi}\,\underline{u}_{\phi} + E_{\theta}\,\underline{u}_{\theta} \tag{3.7}$$

with the transverse field components E_{ϕ} and E_{θ} expressed as follows

$$E_{\phi} = \|\underline{E}\|\sin(\gamma)\,\mathrm{e}^{j\eta} \tag{3.8}$$

$$E_{\theta} = \|\underline{E}\|\cos(\gamma) \tag{3.9}$$

Upon transforming the coordinate system from Spherical to Cartesian, that is, with reference to the unit vectors $\underline{u}_x$, $\underline{u}_y$ and $\underline{u}_z$, the electric field components emerged at the polarisation-sensitive sensor becomes

$$\begin{aligned}\underline{E} = \|\underline{E}\|\Big[&\big(-\sin(\theta)\cos(\gamma) - \cos(\theta)\sin(\phi)\sin(\gamma)\,\mathrm{e}^{j\eta}\big)\underline{u}_x \\ &+ \big(\cos(\theta)\cos(\gamma) - \sin(\theta)\sin(\phi)\sin(\gamma)\,\mathrm{e}^{j\eta}\big)\underline{u}_y \\ &+ \big(\cos(\phi)\sin(\gamma)\,\mathrm{e}^{j\eta}\big)\underline{u}_z\Big]\end{aligned} \tag{3.10}$$

3.2.2 Polarisation-Sensitive Array Response

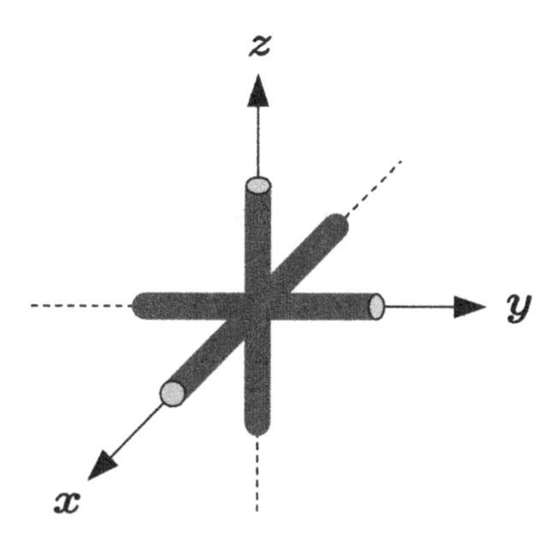

Figure 3.4: A polarisation-sensitive tripole sensor.

Consider a tripole sensor, as illustrated in Figure 3.4, as being the polarisation-sensitive antenna element located at the coordinate origin. The sensor is made up of a set of three co-centered orthogonal dipoles with each of the dipoles measuring and processing each individual part of the signal separately [78]. In order to characterise the response of the sensor, let's define V_x, V_y and V_z as the voltages induced at each of the dipoles as a result of an incoming unit electric field polarised entirely in the x, y and z directions respectively. Suppose a given signal path of arbitrary elliptical polarisation [79] propagates towards an antenna array consisting of N tripole sensors. With the use of Equation (3.10), its associated polarisation-sensitive manifold vector [75] can be extended from Equation (2.4) to form

$$\underline{A}(\underline{\Theta}) = \underline{S}(\theta, \phi) \otimes \underline{\mathrm{q}}(\underline{\Theta}) \tag{3.11}$$

where $\underline{\Theta} \triangleq [\theta, \phi, \gamma, \eta]^T$ is the path's propagation state, and $\underline{\mathrm{q}}(\underline{\Theta}) = [E_x, E_y, E_z]^T$ is a 3-dimensional column vector containing the electric field components induced on each dipole, that is given by

$$\underline{\mathrm{q}}(\underline{\Theta}) = \mathrm{diag}(\underline{\mathcal{V}}) \, . \, \mathbb{T}(\theta, \phi) \, . \, \underline{p}(\gamma, \eta) \tag{3.12}$$

where $\underline{\mathcal{V}} = [\mathrm{V}_x, \mathrm{V}_y, \mathrm{V}_z]^T$ is the sensor's sensitivities to unit electric fields each solely polarised in the x, y, and z directions respectively, $\mathbb{T}(\theta, \phi) = \left[\{1/\cos(\phi)\}.\partial\underline{u}/\partial\theta, \ \partial\underline{u}/\partial\phi\right]$ is the Spherical-to-Cartesian transformation matrix, and $\underline{p} \triangleq \underline{p}(\gamma, \eta) = [\cos(\gamma), \ \sin(\gamma)\, e^{j\eta}]^T$ is the signal's state of polarisation which can be envisioned by making use of the Poincaré sphere concept in Figure 3.3.

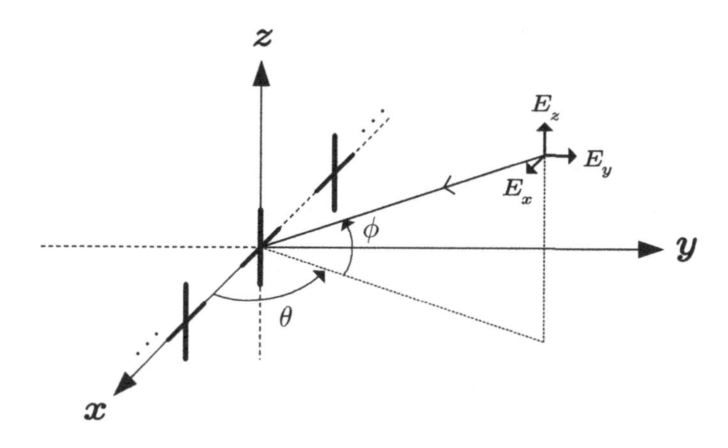

Figure 3.5: A linear crossed-dipole array.

But for a crossed-dipole array as depicted in Figure 3.5, the expression in Equation (3.12) is reduced to $\underline{q}(\Theta) = [E_x, E_z]^T$ consisting of only the field components induced on the horizontally and vertically polarised dipoles respectively. By assuming unity in sensor's sensitivities (i.e. $V_x = V_z = 1$) and that the sensors and the sources are coplanar (i.e. $\phi = 0$), the manifold vector from Equation (3.11) thus becomes

$$\begin{aligned} \underline{A}(\theta, \gamma, \eta) &= \underline{S}(\theta) \otimes \underline{q}(\theta, \gamma, \eta) \\ &= \mathbb{M}(\theta)\, \underline{p}(\gamma, \eta) \end{aligned} \tag{3.13}$$

where $\mathbb{M} \triangleq \mathbb{M}(\theta) = \left[\underline{S}_h(\theta),\ \underline{S}_v(\theta)\right]$ has columns that may be viewed respectively as the spatial array manifold vector for horizontally and vertically polarised signal arriving from the azimuth direction θ.

For some arrays that do not measure and process each polarisation component separately (as in the employment of circularly-polarised sensors in [64]), the above framework in Equation (3.13) can still be utilised by redefining its corresponding polarisation-sensitive manifold vector [74] as follows

$$\underline{A}(\Theta) = \underline{S}(\theta, \phi) \odot \underline{q}(\Theta) \tag{3.14}$$

where $\odot$ is the Hadamard product, and $\underline{q}(\Theta)$ is given by

$$\underline{q}(\Theta) = \mathbb{V}^T . \mathbb{T}(\theta, \phi) . \underline{p}(\gamma, \eta) \tag{3.15}$$

where $\mathbb{V}$ is a $3 \times N$ matrix with its nth column $\left[V_x^{(n)}, V_y^{(n)}, V_z^{(n)}\right]^T$ representing the complex voltages induced at the nth sensor output in response to incoming unit electric fields each polarised solely in the x, y, and z directions respectively.

3.2.3 Diversely-Polarised Manifold Shape

To gain further insights into the general framework derived in Equation (3.13), it is worth looking at the locus circumscribed by the manifold vectors associated with a diversely-polarised array. According to Schmidt [59], there is a distinct difference between the array manifold of a polarisation-sensitive array and that of a polarisation-insensitive array. In either case, its array manifold is a continuous function of the azimuth direction θ. But its distinction arises from the fact that in the polarisation-insensitive case, its array manifold follows a vector continuum as shown in Figure 2.3; whereas in the polarisation-sensitive case, its array manifold forms a bivector continuum as illustrated below.

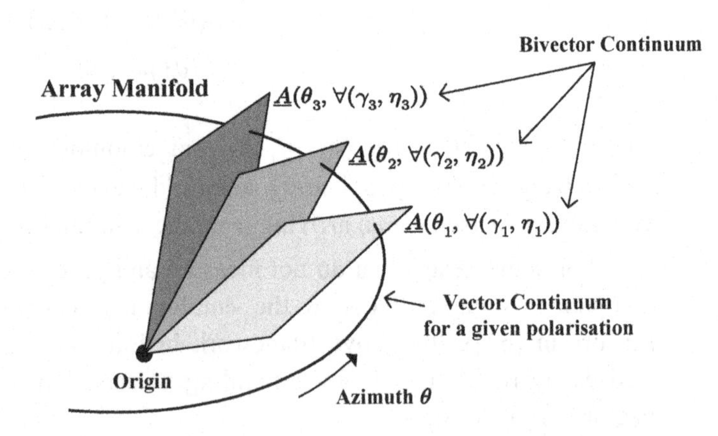

Figure 3.6: Array manifold of polarisation-sensitive antenna array (bivector continuum).

As seen in the figure, the polarisation parameter constitutes a linear parameter subspace. In other words, if a source at a certain azimuth θ is to stay put while transmitting one polarisation after another, its resulting manifold vectors will be confined to a two-dimensional subspace unique to that particular source. Its polarisation state thence determines where in the subspace plane it is located. This continuum of planes, each corresponding to a specific θ, thus brings about the bivector continuum. Mathematically, this allows us to characterise, at every θ direction, the entire range of the source polarisations by using only two orthogonally-polarised manifold vectors lying within the two-dimensional subspace plane as established in Equation (3.13).

3.3 Polar-STAR Signal Modelling

The parameter that plays a significant role in the construction of the Polar-STAR signal model is the polarisation-sensitive crossed-dipole array manifold vector $\underline{A}(\theta, \gamma, \eta)$ as stated in Equation (3.13). The manifold vector stipulates both the spatial and polarimetric dimensions of the signal path. In order to include this polarimetric dimension within the continuous-time signal model, the polarisation-insensitive array manifold vector, employed in the Vector Channel Impulse Response model as described in Section 2.3.2, has to be replaced by its corresponding polarisation-sensitive array manifold vector. With that substitution defined as $\underline{A}_{ij} \triangleq \underline{A}(\theta_{ij}, \gamma_{ij}, \eta_{ij})$ for the jth path of the ith user, the complex baseband received signal vector can therefore be modified from Equation (2.7) to obtain

$$\underline{x}(t) \quad = \quad \sum_{i=1}^{M} \mathbb{A}_i \operatorname{diag}(\underline{\beta}_i) \underline{m}_i(t) + \underline{\mathrm{n}}(t) \tag{3.16}$$

where $\underline{\mathrm{n}}(t)$ is a $2N \times 1$ complex AWGN noise vector appearing across the N pairs of horizontal and vertical dipoles, and

$$\begin{aligned}
\mathbb{A}_i &= \left[\underline{A}_{i1}, \underline{A}_{i2}, \ldots, \underline{A}_{iK_i}\right] && \in \mathcal{C}^{2N \times K_i} \\
\underline{\beta}_i &= \left[\beta_{i1}, \beta_{i2}, \ldots, \beta_{iK_i}\right]^T && \in \mathcal{C}^{K_i \times 1} \\
\underline{m}_i(t) &= \left[m_i(t-\tau_{i1}), m_i(t-\tau_{i2}), \ldots, m_i(t-\tau_{iK_i})\right]^T && \in \mathcal{C}^{K_i \times 1}
\end{aligned}$$

In a similar manner, the spatial array manifold vector, used in the STAR manifold formulation developed in Equation (2.16), can be replaced with the crossed-dipole array manifold vector to form the following Polar-STAR manifold expression

$$\underline{\mathfrak{h}}_{ij} \triangleq \underline{A}_{ij} \otimes \mathbb{J}^{l_{ij}} \underline{\mathfrak{c}}_i \tag{3.17}$$

Notice that the above manifold vector has effectively brought together the spatial, temporal, as well as the polarimetric, elements of the vector channel composition. Now by denoting $\mathbb{H}_i = \left[\underline{\mathfrak{h}}_{i1}, \underline{\mathfrak{h}}_{i2}, \ldots, \underline{\mathfrak{h}}_{iK_i}\right]$ as the Polar-STAR manifold vectors associated with all the K_i multipaths of the ith user, the net polarimetric-spatial-temporal discretised signal vector $\underline{x}[n]$ received by the crossed-dipole array can thus be written as

$$\underline{x}[n] = \sum_{i=1}^{M} \overline{\mathbb{H}}_i \mathbb{G}_i \underline{\mathrm{a}}_i[n] + \underline{\mathrm{n}}[n] \tag{3.18}$$

where $\overline{\mathbb{H}}_i = \left[(\mathbb{I}_{2N} \otimes (\mathbb{J}^T)^{L/2})\mathbb{H}_i, \mathbb{H}_i, (\mathbb{I}_{2N} \otimes \mathbb{J}^{L/2})\mathbb{H}_i\right]$ contains the channel matrices, $\mathbb{G}_i = \mathbb{I}_3 \otimes \underline{\beta}_i$ constitutes the complex multipath coefficients, and $\underline{\mathrm{a}}_i[n] = \left[\mathrm{a}_i[n-1], \mathrm{a}_i[n], \mathrm{a}_i[n+1]\right]^T$ incorporates the previous, current and next symbol contributions. The inclusion of the polarimetric dimension in the above discretised signal vector aids in the exploitation of the joint spatial-temporal structure in twofolds. It offers (1) an extra degree of multipath differentiation and (2) an additional domain of signal diversity, in the channel estimation and reception architecture. However note that the main focus of this work is to devise a parametric estimation algorithm for the polarimetric-spatial-temporal parameters, but not for the multipath coefficient which can nevertheless be obtained by applying for instance [50] after performing the polarisation-space-time estimation.

3.4 Vector Channel Estimation and Reception

Specifically-designed vector channel estimation and reception algorithm is devised in this section to exploit the polarisation-space-time dimensions inherent in the environment. A preprocessing operation is first performed onto the received discretised signal vector to facilitate the implementation of the channel estimation procedure. This is then followed by an extraction of the polarimetric, spatial and temporal channel parameters which are to be fed into the reception process to support a recovery of the desired user's transmitted signal from the multiple-users' multipath propagation environment.

3.4.1 Operation of Preprocessor

In order to isolate the desired signal component in Equation (3.18), the discretised signal vector $\underline{x}[n]$ is passed through the desired $\mathcal{D}^{\text{th}}$ user's preprocessor which is given by

$$\mathbb{Z}_{\mathcal{D}} = \mathbb{I}_{2N} \otimes \left(\text{diag}(\underline{\tilde{c}}_{\mathcal{D}})^{-1}\mathbb{F}\right) \quad (3.19)$$

where $\underline{\tilde{c}}_{\mathcal{D}}$ is the Fourier transformed version of the desired user's reference code vector $\underline{c}_{\mathcal{D}}$, that is $\underline{\tilde{c}}_{\mathcal{D}} = \mathbb{F}\underline{c}_{\mathcal{D}}$ with $\mathbb{F}$ being the $L \times L$ Fourier transformation matrix defined as

$$\mathbb{F} = \begin{bmatrix} 1 & 1 & 1 & \cdots & 1 \\ 1 & \Phi^1 & \Phi^2 & \cdots & \Phi^{(L-1)} \\ 1 & \Phi^2 & \Phi^4 & \cdots & \Phi^{2(L-1)} \\ \vdots & \vdots & \vdots & \ddots & \vdots \\ 1 & \Phi^{(L-1)} & \Phi^{2(L-1)} & \cdots & \Phi^{(L-1)^2} \end{bmatrix} \quad (3.20)$$

$$= \left[\underline{\Phi}^0,\ \underline{\Phi}^1,\ \underline{\Phi}^2,\ \ldots,\ \underline{\Phi}^{(L-1)}\right]$$

where $\underline{\Phi} = \left[1,\ \Phi^1,\ \Phi^2,\ \ldots,\ \Phi^{(L-1)}\right]^T$

with $\Phi = \exp\left(-j\dfrac{2\pi}{L}\right)$

The underlying principle behind the construction of the preprocessor is based on the decoupling property provided by the application of the Fourier transformation procedure. To see this, consider below the Fourier transformed version of the temporally-delayed code vector previously introduced in Equation (2.11):

$$\begin{aligned} \underline{\tilde{\mathfrak{c}}}_{ij} &= \mathbb{F}\left(\underline{\mathfrak{c}}_{ij}\right) \\ &= \mathbb{F}\left(\mathbb{J}^{l_{ij}}\underline{\mathfrak{c}}_{i}\right) \\ &= \text{diag}(\underline{\tilde{\mathfrak{c}}}_{i})\,\underline{\Phi}^{l_{ij}} \end{aligned} \tag{3.21}$$

As a result of the Fourier transformation operation, the delayed form of the sampled spreading waveform $\underline{\mathfrak{c}}_{ij}$ for every j^{th} path of the ith user is effectively decoupled into two terms: (1) the Fourier transformed version of the ith user's zero-delayed reference code signature $\underline{\mathfrak{c}}_i$, and (2) the discretised path delay l_{ij} which is now separated and independent of the sampled spreading code vector. The transformation has, in essence, turned the time-domain path delay into a frequency-domain complex exponential term $\underline{\Phi}^{l_{ij}}$. Its path delay vector representation from the frequency-domain has a Vandermonde geometric progression structure which is similar to that of the array manifold vector representation from a uniform linear array (ULA) configuration. The only main difference is that in the case of the array manifold vector, its unknown parameter to be estimated is the signal path's Direction-of-Arrival (DOA); whereas in the case of the complex exponential vector, this unknown parameter is the signal path's Time-of-Arrival (TOA).

To appreciate the effect of this preprocessing operation on the discretised signal vector $\underline{x}[n]$, let's now apply the preprocessor to the Polar-STAR manifold vector in Equation (3.17) as follows

$$\begin{aligned} \underline{\tilde{\mathfrak{h}}}_{ij} &= \mathbb{Z}_{\mathcal{D}}\,\underline{\mathfrak{h}}_{ij} \\ &= \left\{\mathbb{I}_{2N} \otimes \left(\text{diag}(\underline{\tilde{\mathfrak{c}}}_{\mathcal{D}})^{-1}\mathbb{F}\right)\right\}.\left(\underline{A}_{ij} \otimes \mathbb{J}^{l_{ij}}\underline{\mathfrak{c}}_{i}\right) \\ &= \underline{A}_{ij} \otimes \left\{\left(\text{diag}(\underline{\tilde{\mathfrak{c}}}_{\mathcal{D}})^{-1}\mathbb{F}\right).\left(\mathbb{J}^{l_{ij}}\underline{\mathfrak{c}}_{i}\right)\right\} \\ &= \underline{A}_{ij} \otimes \left\{\text{diag}(\underline{\tilde{\mathfrak{c}}}_{\mathcal{D}})^{-1}\text{diag}(\underline{\tilde{\mathfrak{c}}}_{i})\,\underline{\Phi}^{l_{ij}}\right\} \end{aligned} \tag{3.22}$$

Notice that the above expression can be reduced to simply $\underline{A}_{\mathcal{D}j} \otimes \underline{\Phi}^{l_{\mathcal{D}j}}$ if and only if $i = \mathcal{D}$ which corresponds to the desired user of interest. Hence by applying the same operation onto Equation (3.18), the net received discretised signal vector is thus transformed to

$$\begin{aligned} \underline{y}[n] &= \mathbb{Z}_{\mathcal{D}}\, \underline{x}[n] \\ &= \tilde{\mathbb{H}}_{\mathcal{D}}\, \underline{\beta}_{\mathcal{D}} \mathrm{a}_{\mathcal{D}}[n] + \mathbb{Z}_{\mathcal{D}}\, \underline{I}_{\mathrm{ISI}}[n] + \mathbb{Z}_{\mathcal{D}} \underline{I}_{\mathrm{MAI}}[n] + \mathbb{Z}_{\mathcal{D}}\, \mathrm{n}[n] \qquad (3.23) \end{aligned}$$

where $\tilde{\mathbb{H}}_{\mathcal{D}} = \left[\underline{A}_{\mathcal{D}1} \otimes \underline{\Phi}^{l_{\mathcal{D}1}},\ \underline{A}_{\mathcal{D}2} \otimes \underline{\Phi}^{l_{\mathcal{D}2}},\ \ldots,\ \underline{A}_{\mathcal{D}K_{\mathcal{D}}} \otimes \underline{\Phi}^{l_{\mathcal{D}K_{\mathcal{D}}}}\right]$. Note that the preprocessed signal vector $\underline{y}[n]$ is decomposed into its four constituents, namely the desired signal term $\tilde{\mathbb{H}}_{\mathcal{D}} \underline{\beta}_{\mathcal{D}} \mathrm{a}_{\mathcal{D}}[n]$, the ISI, MAI and noise components respectively.

3.4.2 Spatial-Temporal Smoothing Technique

However the second order statistics of the signal vector $\underline{y}[n]$, instead of providing a basis for the desired signal subspace, would result in a rank deficiency with the desired signal subspace dimension being reduced to one. This is because, as shown in Equation (3.23), the columns of the matrix $\tilde{\mathbb{H}}_{\mathcal{D}}$ are linearly combined by the path coefficient vector $\underline{\beta}_{\mathcal{D}}$; hence its contribution to the observation space of $\underline{y}[n]$ will consequently lead to a subspace of only one dimension. To restore the dimensionality of this subspace back to $K_{\mathcal{D}}$, the Vandermonde structure of the submatrices of $\tilde{\mathbb{H}}_{\mathcal{D}}$ provided by the preprocessing operation will be exploited. This can be achieved by performing a technique referred to as temporal smoothing, a concept similar to that of spatial smoothing described in [80].

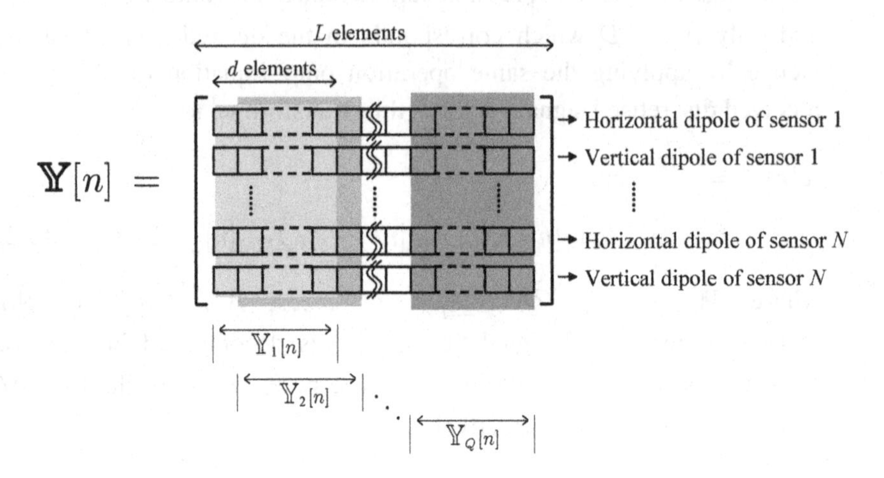

Figure 3.7: Temporal smoothing technique.

For clarity, let's reshape the preprocessed signal vector $\underline{y}[n] \in \mathcal{C}^{2NL\times 1}$ to a matrix $\mathbb{Y}[n] \in \mathcal{C}^{2N\times L}$ so that every of its successive two rows reflect the preprocessed output for each crossed-dipole sensor, as illustrated in Figure 3.7. By extracting a set of Q (where $Q = L - d + 1$) overlapping submatrices of length d (where $d < L$), and concatenating each of the submatrices via vectorisation, Q concatenated subvectors each of length Nd are thus formed, i.e.

$$\underline{y}_v[n] \quad = \quad \text{vec}(\mathbb{Y}_v[n]), \qquad \forall v = 1, 2, \ldots, Q \tag{3.24}$$

where vec(•) is the row-wise vectorisation operator. With that, the $2Nd \times 2Nd$ temporal-smoothed covariance matrix $\mathbb{R}_{\text{Tsmooth}}$ can therefore be obtained as follows

$$\mathbb{R}_{\text{Tsmooth}} \quad = \quad \frac{1}{Q}\sum_{v=1}^{Q} \mathbb{R}_{\underline{y}_v} \tag{3.25}$$

where $\mathbb{R}_{\underline{y}_v}$ is the covariance matrix obtained from the subvector $\underline{y}_v[n]$. Note that the dimension can only be successfully restored to $K_{\mathcal{D}}$, provided that $Q \geq K_{\mathcal{D}}$. This technique also applies for paths arriving from the same direction (co-directional). However, for paths arriving at the same time (co-delay), singularity in $\mathbb{R}_{\text{Tsmooth}}$ will occur. This special case cannot be resolved for a general array geometry but for a uniform linear array where spatial smoothing can be performed on the sensor array [80] as pictorially represented in Figure 3.8.

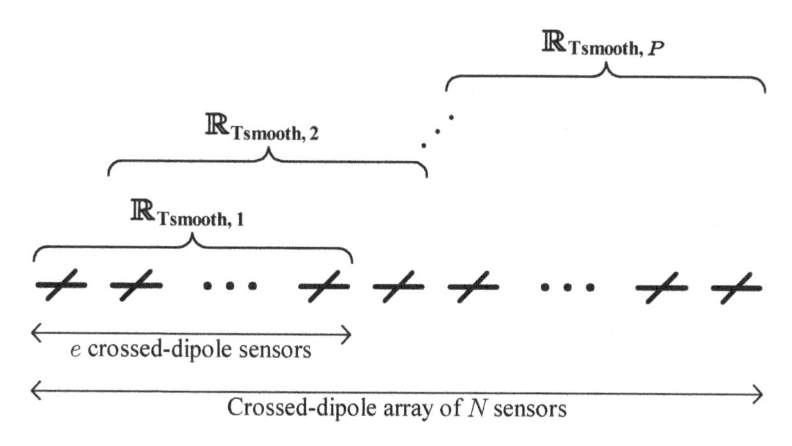

Figure 3.8: Spatial smoothing technique.

The idea is to divide the crossed-dipole array, consisting of N horizontal and vertical dipoles pairs, into P overlapping subarrays with each containing $e = N - P + 1$ crossed-dipole sensors. The temporal-smoothed covariance matrices associated with each and every of the overlapping subarrays are then averaged out to form the spatial-temporal-smoothed covariance matrix $\mathbb{R}_{\text{STsmooth}}$ as shown below

$$\mathbb{R}_{\text{STsmooth}} = \frac{1}{P} \sum_{u=1}^{P} \mathbb{R}_{\text{Tsmooth}, u} \tag{3.26}$$

3.4.3 Polarisation-Angle-Delay Estimation (PADE) Algorithm

A. Complete Normalised Polar-STAR cost function

From the above discussion, a MUSIC-type cost function, utilising the preprocessed Polar-STAR manifold vector, can be shown to be based on the following criterion:

$$\xi(\theta, l, \gamma, \eta) = \frac{(\underline{A} \otimes \underline{\Phi}_d^l)^H \mathbb{E}_n \mathbb{E}_n^H (\underline{A} \otimes \underline{\Phi}_d^l)}{(\underline{A} \otimes \underline{\Phi}_d^l)^H (\underline{A} \otimes \underline{\Phi}_d^l)}$$

$$= \frac{\underline{p}^H (\mathbb{M} \otimes \underline{\Phi}_d^l)^H \mathbb{E}_n \mathbb{E}_n^H (\mathbb{M} \otimes \underline{\Phi}_d^l) \underline{p}}{\underline{p}^H (\mathbb{M} \otimes \underline{\Phi}_d^l)^H (\mathbb{M} \otimes \underline{\Phi}_d^l) \underline{p}} \qquad (3.27)$$

where $\underline{\Phi}_d$ is a subvector of $\underline{\Phi}$ with length d, and $\mathbb{E}_n$ is a matrix whose columns are the generalised noise eigenvectors of $(\mathbb{R}_{\text{STsmooth}}, \mathbb{D})$ due to the transformed noise in Equation (3.23), with $\mathbb{D}$ representing the spatial-temporal-smoothed diagonal matrix $\mathbb{Z}_{\mathcal{D}} \mathbb{Z}_{\mathcal{D}}^H$. However the minimisation process involves a multidimensional search over θ, l, γ and η for its minima. To simplify the process, it is noted that the search over the polarisation space $\underline{p}$ is equivalent to satisfying the following minimum generalised eigenvector $\underline{\text{k}}_{\text{min}}$ and eigenvalue λ_{min}:

$$(\mathbb{M} \otimes \underline{\Phi}_d^l)^H \mathbb{E}_n \mathbb{E}_n^H (\mathbb{M} \otimes \underline{\Phi}_d^l) \underline{\text{k}}_{\text{min}} = \lambda_{\text{min}} (\mathbb{M} \otimes \underline{\Phi}_d^l)^H (\mathbb{M} \otimes \underline{\Phi}_d^l) \underline{\text{k}}_{\text{min}} \qquad (3.28)$$

Hence by applying the quadratic formula and denoting det(•) as the determinant of a 2 × 2 matrix, the cost function in Equation (3.27) can be simplified to

$$\xi(\theta, l) = \left(\zeta - \sqrt{\zeta^2 - 4\det(\mathbb{C})\det(\mathbb{B})}\right) \Big/ \det(\mathbb{B}) \qquad (3.29)$$

where

$$\mathbb{C} \triangleq \begin{bmatrix} c_{11} & c_{12} \\ c_{21} & c_{22} \end{bmatrix} = (\mathbb{M} \otimes \underline{\Phi}_d^l)^H \mathbb{E}_n \mathbb{E}_n^H (\mathbb{M} \otimes \underline{\Phi}_d^l) \qquad (3.30)$$

$$\mathbb{B} \triangleq \begin{bmatrix} b_{11} & b_{12} \\ b_{21} & b_{22} \end{bmatrix} = (\mathbb{M} \otimes \underline{\Phi}_d^l)^H (\mathbb{M} \otimes \underline{\Phi}_d^l) \qquad (3.31)$$

and $\zeta = (c_{11} b_{22} + c_{22} b_{11}) - (c_{12} b_{21} + c_{21} b_{12})$. Now let ξ_{min} be the minima obtained from the spectrum constructed using the cost function $\xi(\theta, l)$ in Equation (3.29). The location of the minima, as such, provides the joint estimate of its Direction-of-Arrival (DOA) and Time-of-Arrival (TOA). Its corresponding polarisation parameters, if required, can be estimated as follows:

$$\begin{aligned} \widehat{\gamma} &= \tan^{-1}(|\rho|) \\ \widehat{\eta} &= \arg(\rho/|\rho|) \end{aligned} \tag{3.32}$$

$$\text{where } \rho = \begin{cases} (\xi_{\min} b_{11} - 2c_{11})/(2c_{12} - \xi_{\min} b_{12}) \\ \text{or} \\ (\xi_{\min} b_{21} - 2c_{21})/(2c_{22} - \xi_{\min} b_{22}) \end{cases}$$

It is worthwhile to note that the above expression ρ, if necessary, can be further simplified in most circumstances since the value $\xi_{\min}$ is usually close to zero.

B. Compact Un-Normalised Polar-STAR cost function

Instead of applying the above normalised algorithmic function, a more compact optimisation algorithm can be devised by considering the un-normalised form of the subspace-based cost function as shown below

$$\begin{aligned} \xi(\theta, l, \gamma, \eta) &= (\underline{A} \otimes \underline{\Phi}_d^l)^H \mathbb{E}_n \mathbb{E}_n^H (\underline{A} \otimes \underline{\Phi}_d^l) \\ &= \underline{p}^H (\mathbb{M} \otimes \underline{\Phi}_d^l)^H \mathbb{E}_n \mathbb{E}_n^H (\mathbb{M} \otimes \underline{\Phi}_d^l) \underline{p} \end{aligned} \tag{3.33}$$

Likewise the operation involves an exhaustive multidimensional search across θ, l, γ and η. A more efficient minimisation function can be found by adopting the 2×2 matrix $\mathbb{C}$ as shown in Equation (3.30). Notice that the minimisation over the polarisation space of $\underline{p}$ is equivalent to finding the eigenvectors corresponding to the minimum eigenvalues of $\mathbb{C}$. Hence by dropping the two-dimensional vector $\underline{p}$ and applying the quadratic formula, the cost function in Equation (3.33) can be reduced to become

$$\xi(\theta, l) = \text{trace}(\mathbb{C}) - \sqrt{\text{trace}(\mathbb{C})^2 - 4\det(\mathbb{C})} \tag{3.34}$$

where trace(•) denotes the trace operation of a 2×2 matrix. Similarly, the location of the minima will yield the joint space-time estimates; the value of the minima $\xi_{\min}$, on the other hand, will be employed to generate its corresponding polarisation parameters:

$$\begin{aligned} \widehat{\gamma} &= \tan^{-1}(|\rho|) \\ \widehat{\eta} &= \arg(\rho/|\rho|) \end{aligned} \qquad (3.35)$$

$$\text{where } \rho = \begin{cases} (\xi_{\min} - 2c_{11})/2c_{12} \\ \text{or} \\ 2c_{21}/(\xi_{\min} - 2c_{22}) \end{cases}$$

Again further simplification can be carried out, if necessary, on the above expression ρ since the value $\xi_{\min}$ is normally insignificant.

Remark: Note that the un-normalised cost function, unlike its normalised counterpart, is only applicable for arrays that employed sensor elements that are insensitive to the arriving angles of the signals [74]. In other words, each of the sensor elements has a constant gain over a specified spatial field-range of interest [81], as in the case of a disk cone sensor. However, for sensor elements that are directionally-sensitive, such as the above crossed-dipole sensors, its matrix $\mathbb{M}$ will contain an unwanted trigonometric term which is dependent on the DOA of the incoming signal. This will consequently affect the consistency in the performance of the space-time spectral analysis with respect to its spatial searching domain since its subspace-based cost function is un-normalised (its normalised counterpart, in contrast, will not exhibit such performance inconsistency [75]). To resolve this with the compact un-normalised function, this DOA-dependent term can be decoupled from the matrix $\mathbb{M}$ and lumped together with the vector $\underline{p}$ of Equation (3.13). Take for instance the crossed-dipole array configuration in Figure 3.5. Upon decoupling, the vector $\underline{p}(\gamma, \eta)$ will become $\underline{p}(\theta, \gamma, \eta) = [\sin(\theta)\cos(\gamma), \sin(\gamma)\, e^{j\eta}]^T$ with $\sin^2(\theta)$ being proportional to the power received at each of the sensor elements with respect to the arriving angle of the signal. This $\sin^2(\theta)$ term will be in unity when the sensor elements are insensitive to the DOA of the incoming signal. But note that as a result of such decoupling, the computation of ρ in Equation (3.35) will have to include a multiplicative term $\sin(\widehat{\theta})$ where $\widehat{\theta}$ is the estimated DOA [76], given as $\rho = (\xi_{\min} - 2c_{11})\sin(\widehat{\theta})/2c_{12}$ or $\rho = 2c_{21}\sin(\widehat{\theta})/(\xi_{\min} - 2c_{22})$.

Consider an ill-conditioned situation in which two or more multipaths of differing polarisation are closely located in space and time, such that their minima are indistinguishable or unresolvable from one another in the space and time domain. In such an event, the eigenvalues of the matrix $\mathbb{C}$ would be approximately near to zero, thus degenerating $\mathbb{C}$ to a near trivial matrix. By exploiting the supplementary eigenvalue of the matrix $\mathbb{C}$, the following cost function can hence be utilised to detect such occurrences.

$$\xi(\theta, l) = \text{trace}(\mathbb{C}) + \sqrt{\text{trace}(\mathbb{C})^2 - 4\det(\mathbb{C})} \qquad (3.36)$$

3.4.4 Polarisation-Space-Time Reception

Having performed the polarisation-space-time estimation using the devised PADE algorithm described in the previous section, the Polar-STAR manifold vectors associated with the desired user can be reconstructed as follows

$$\widehat{\mathbb{H}}_{\mathcal{D}} = \left[\underline{\widehat{\mathfrak{h}}}_{\mathcal{D}1}, \ \underline{\widehat{\mathfrak{h}}}_{\mathcal{D}2}, \ \ldots, \ \underline{\widehat{\mathfrak{h}}}_{\mathcal{D}K_{\mathcal{D}}} \right] \qquad (3.37)$$

Now by suppressing the contributions due to its polarimetric ISI and MAI interferences from the received discretised signal vector $\underline{x}[n]$ in Equation (3.18), the decision statistic for the nth symbol of the desired user can thus be obtained as

$$b[n] = \underline{w}^H . \underline{x}[n] \qquad (3.38)$$

where $\underline{w}$ is the desired user's weight vector based on the projection of the noise subspace spanned by its interferences, i.e.

$$\underline{w} = \mathbb{P}_{\mathbb{E}_\mathrm{n}} \widehat{\mathbb{H}}_{\mathcal{D}} \left(\widehat{\mathbb{H}}_{\mathcal{D}}^H \, \mathbb{P}_\mathrm{n} \widehat{\mathbb{H}}_{\mathcal{D}} \right)^{-1} \underline{\widehat{\beta}}_{\mathcal{D}} \qquad (3.39)$$

where $\mathbb{P}_{\mathbb{E}_\mathrm{n}} = \mathbb{E}_\mathrm{n} (\mathbb{E}_\mathrm{n}^H \, \mathbb{E}_\mathrm{n})^{-1} \mathbb{E}_\mathrm{n}^H$ is the projection matrix onto the subspace spanned by the noise eigenvectors of $\mathbb{R}_{intf} = \mathbb{R}_{xx} - \widehat{\mathbb{H}}_{\mathcal{D}} \underline{\widehat{\beta}}_{\mathcal{D}} \underline{\widehat{\beta}}_{\mathcal{D}}^H \widehat{\mathbb{H}}_{\mathcal{D}}^H$ in which $\mathbb{R}_{xx}$ is the covariance matrix of Equation (3.18).

The complete procedure outlining the major steps of the proposed Polar-STAR receiver, with its front-end integrated with the polarisation-angle-delay channel estimator, is summarised as shown below:

1. Sample the crossed-dipole array output and concatenate its tapped-delay lines (TDL) contents to form the discretised signal vector $\underline{x}[n]$.

2. Apply the desired user's preprocessor as shown in Equation (3.19) onto the discretised signal vector to obtain its transformed signal vector $\underline{y}[n]$. Next perform the temporal smoothing technique to restore the dimensionality of the desired signal subspace. Note that for paths arriving at the same time, spatial smoothing can be performed on top of temporal smoothing to construct the spatial-temporal-smoothed covariance matrix $\mathbb{R}_{\text{STsmooth}}$.

3. Evaluate the compact un-normalised cost function in Equation (3.34), based on the eigenvector decomposition of $\mathbb{R}_{\text{STsmooth}}$, to obtain the joint space-time estimation of the desired user's signal path. Its corresponding polarisation estimates can be found by using the set of analytical equations in Equation (3.35).

   Remark: In an ill-conditioned event whereby two or more closely-located multipaths are spatially and temporally non-differentiable, the supplementary cost function in Equation (3.36) can be employed to detect such occurrences.

4. Compute the desired user's weight vector based on the estimated polarimetric-spatial-temporal parameters as described in Equation (3.39). Finally apply the calculated weight vector onto the received discretised signal vector $\underline{x}[n]$ to realise the nth symbol's decision statistic due to the desired user.

3.5 Simulation Studies

In this section, several illustrative examples are presented to demonstrate the key features of the proposed PADE algorithm. Consider a linear array consisting of $N = 5$ crossed-dipole sensors (with each having half-wavelength spacing) operating in the presence of $M = 3$ co-channel CDMA users, where each user is being assigned a unique Gold sequence of length $\mathcal{N}_c = 31$. The array is assumed to collect 200 data symbols, with chip-rate sampling, for processing.

Table 3.1: Users' parameters

Path	User 1				User 2				User 3			
	θ_{1j}	l_{1j}	γ_{1j}	η_{1j}	θ_{2j}	l_{2j}	γ_{2j}	η_{2j}	θ_{3j}	l_{3j}	γ_{3j}	η_{3j}
$j=1$	40°	$8T_c$	0°	–	50°	$18T_c$	80°	30°	30°	$3T_c$	10°	0°
$j=2$	50°	$20T_c$	90°	–	70°	$10T_c$	10°	-70°	45°	$24T_c$	60°	100°
$j=3$	60°	**$15T_c$**	45°	90°	80°	$5T_c$	40°	120°	60°	$5T_c$	20°	50°
$j=4$	85°	$10T_c$	45°	-90°	**95°**	**$20T_c$**	**50°**	**90°**	80°	$20T_c$	70°	-90°
$j=5$	**100°**	$5T_c$	30°	-10°	**95.5°**	**$20T_c$**	**45°**	**0°**	105°	$11T_c$	35°	-20°
$j=6$	**100°**	$21T_c$	5°	-180°	110°	$6T_c$	70°	100°	120°	$28T_c$	45°	45°
$j=7$	120°	**$15T_c$**	20°	80°	130°	$25T_c$	30°	-160°	150°	$20T_c$	50°	170°
$j=8$	130°	$25T_c$	60°	120°	140°	$15T_c$	20°	5°	160°	$12T_c$	80°	-140°

3.5.1 Performance of Polarisation-Space-Time Algorithm

Assume that user 1 is the desired user, with an input SNR of 20dB, together with 2 other interferers each constituting a SIR of -20dB (i.e. near-far problem). All 3 users are assumed to have 8 multipath signals each, with their parameters as listed in Table 3.1. By partitioning the array into 2 overlapping subarrays (i.e. each having 4 crossed-dipoles) for spatial smoothing and setting $d = 50$ (i.e. $Q = 13$) for temporal smoothing, it can be seen from Figure 3.9 that all the 8 multipaths, as well as the co-delay and co-directional paths, can be identified/estimated successfully using the proposed space-time estimation algorithm. In addition, their corresponding polarisation parameters can also be estimated as shown in Table 3.2 by using the set of analytical equations given in Equation (3.35). Notice that the number of resolvable paths, unlike that of the classical MUSIC algorithm, is no longer limited by the number of sensors available in the array.

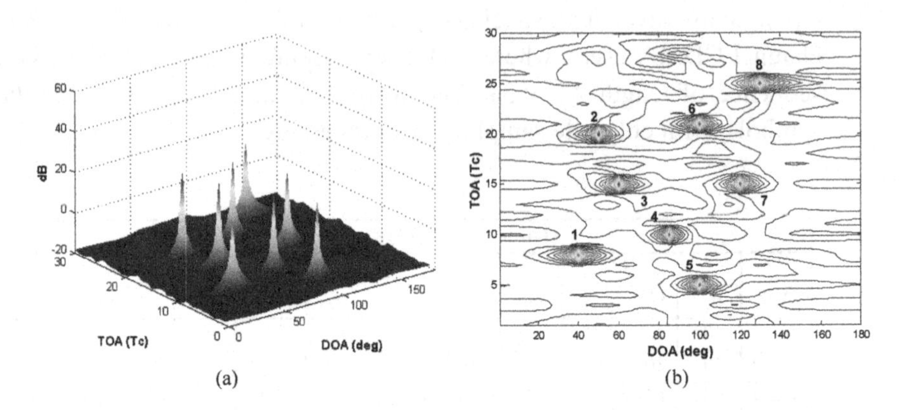

Figure 3.9: Spatial-temporal spectrum showing all 8 multipaths due to desired user.

Table 3.2: Estimated polarisation parameters (User 1)								
Path	$j=1$	$j=2$	$j=3$	$j=4$	$j=5$	$j=6$	$j=7$	$j=8$
$\widehat{\gamma}_{1j}$	0.06°	89.89°	45.03°	44.99°	29.91°	5.06°	19.91°	59.80°
$\widehat{\eta}_{1j}$	–	–	89.91°	-90.09°	-10.08°	-179.25°	79.90°	120.02°

In contrast to that of a linearly polarised array, it is also clear that the crossed-dipole array is able to handle any arbitrary polarised multipath signals, including signal path which is horizontally polarised (path 1), vertically polarised (path 2), left-hand circularly polarised (path 3) and right-hand circularly polarised (path 4). To appreciate this, let's compare the constellation diagram of the proposed Polar-STAR receiver utilising crossed-dipole array with that of a conventional STAR receiver utilising vertically polarised array as illustrated in Figure 3.10. It is thus apparent that by harnessing the diverse polarisations inherent in the multipath signals, the proposed receiver is seen to perform much better than its conventional counterpart.

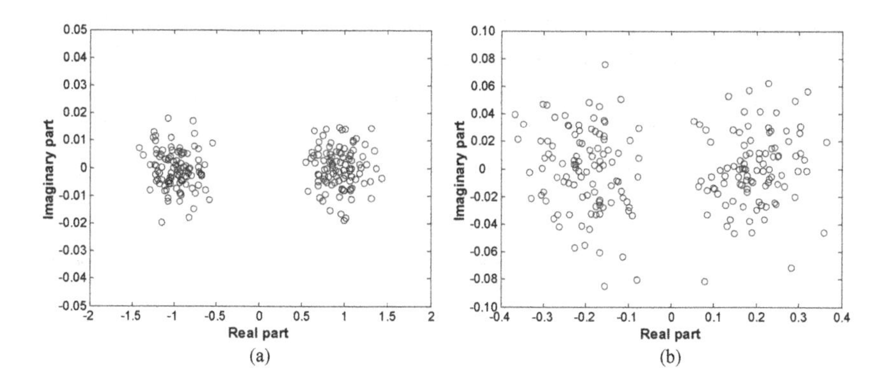

Figure 3.10: Constellation diagrams of (a) proposed Polar-STAR receiver using crossed-dipole array, and (b) conventional STAR receiver using vertically polarised array.

Figure 3.11 depicts the standard deviations of the polarisation, DOA and TOA estimates of the desired user for a normal-incidence desired signal path delayed over $15T_c$ with a SNR of -10dB. It is assumed that the desired user operates in the presence of 4 other interferers (with each interferer having 8 multipaths), individually constituting an interference ratio of 40dB. The deviations are plotted against its polarisation parameters Gamma γ (with Eta $\eta = 0°$) and Eta η (with Gamma $\gamma = 45°$), to demonstrate respectively the performance in response to linearly-polarised state (ranging from horizontal polarisation to vertical polarisation) and elliptically-polarised state (ranging from right-hand circular polarisation to left-hand circular polarisation). All the results of the estimate were averaged over 200 Monte-Carlo simulations. Note that the standard deviation of the polarisation estimate [65] is defined as the angular distance on the Poincaré sphere δ, where $0 \leq \delta \leq \pi$, obtained using

$$\cos(\delta) = \cos(2\gamma)\cos(2\widehat{\gamma}) + \sin(2\gamma)\sin(2\widehat{\gamma})\cos(\eta - \widehat{\eta}) \quad (3.40)$$

From Figure 3.11, it is therefore evidential that despite in a poor SNR environment, with strong interferers constituting the near-far problem effect situation, the proposed algorithm is still able to provide relatively consistent and accurate estimates irrespective of the polarisations of the received signal.

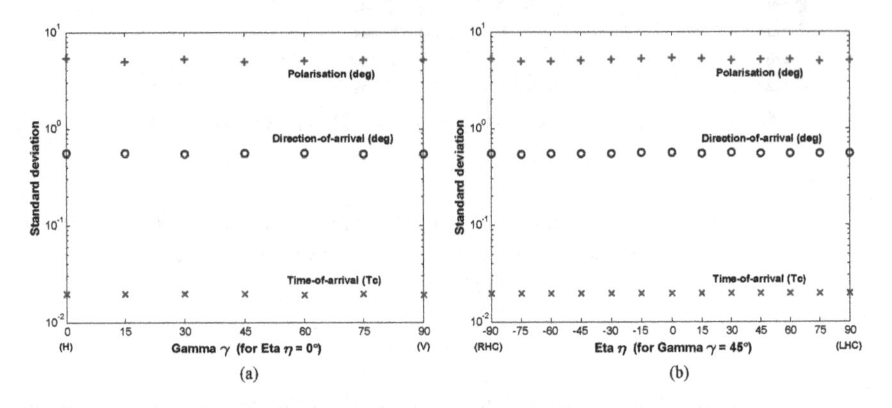

Figure 3.11: Standard deviation of estimates vs (a) Gamma γ with Eta $\eta = 0°$ and (b) Eta η with Gamma $\gamma = 45°$.

Next, let's take a closer look at the effect of polarisation on the performance of a crossed-dipole array as compared with a vertically polarised array and a horizontally polarised array. Consider the case of a single signal path with $\theta = 80°$ and $l = 22T_c$. Assuming an SNR of 0dB and SIR = -30dB, the standard deviation of the estimates with respect to its orientation angle κ (see Equation (3.5)) and ellipticity angle ϵ (see Equation (3.6)) are averaged over 100 Monte Carlo runs and plotted as shown in Figure 3.12. As expected, the estimates deteriorate rapidly when the signal path tends to vertical polarisation ($\kappa \rightarrow 90°$) for a horizontally polarised array; and when the signal path tends to horizontal polarisation ($\kappa \rightarrow 0°$ or $180°$) for a vertically polarised array as illustrated in Figure 3.12a. Similarly from Figure 3.12b, the estimates degenerate when the signal path's elliptical polarisation approaches linear polarisation ($\epsilon \rightarrow 0°$) for a vertically polarised array. On the contrary, the estimates due to a crossed-dipole array have in general a lower and relatively constant standard deviation irregardless of the received signal's polarisation. But note that this deviation will increase as the signal gravitates towards the endfire direction of the array, especially when its polarisation tends horizontal.

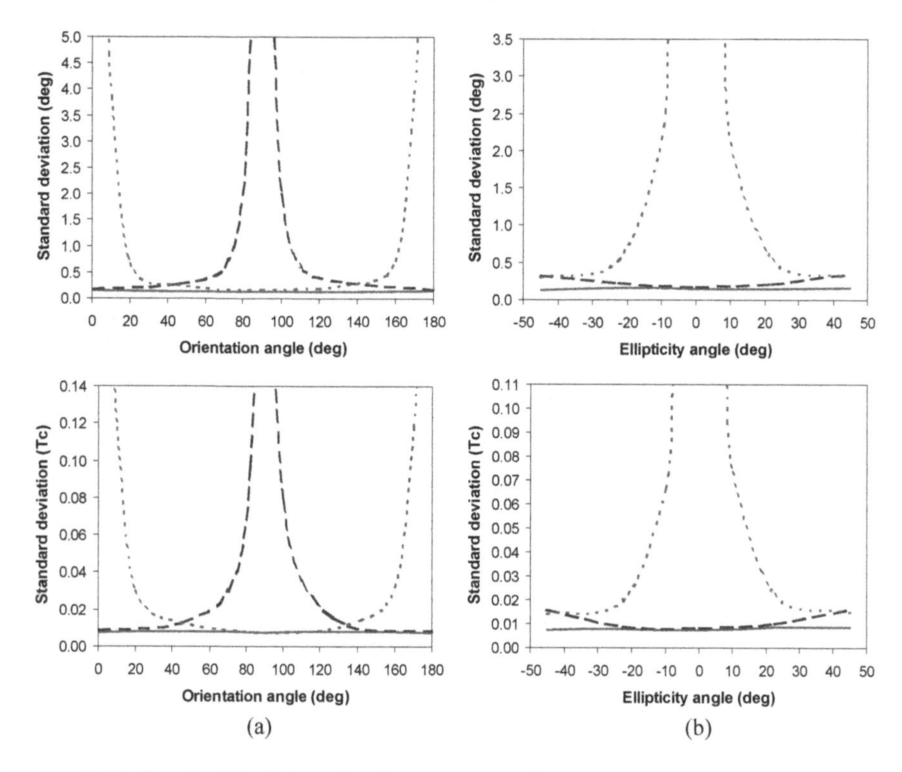

Figure 3.12: Standard deviation of estimates vs (a) orientation angle κ with $\epsilon = 0°$ (linear polarisation) & (b) ellipticity angle ϵ with $\kappa = 0°$ (elliptical polarisation), using crossed-dipole array (solid line), horizontally-polarised array (dashed line) and vertically-polarised array (dotted line).

3.5.2 Studies of Closely-Located Paths

Now let's compare the performance of a crossed-dipole array with that of an equivalent polarisation-insensitive array, with the latter assumed to be omitting any polarisation mismatches. For simplicity, the desired user 1 is assumed to have only 2 linearly polarised multipaths: $(\theta_{11}, l_{11}, \gamma_{11}, \eta_{11}) = (80^\circ, 10T_c, 20^\circ, 0^\circ)$ and $(\theta_{12}, l_{12}, \gamma_{12}, \eta_{12}) = (82^\circ, 11T_c, \gamma, 0^\circ)$ with their polarisation difference defined as $\Delta\gamma = (\gamma - 20^\circ)$. Using the same scenario as that in Table 3.1, the standard deviation of the estimates associated with the first multipath is plotted as shown in Figure 3.13.

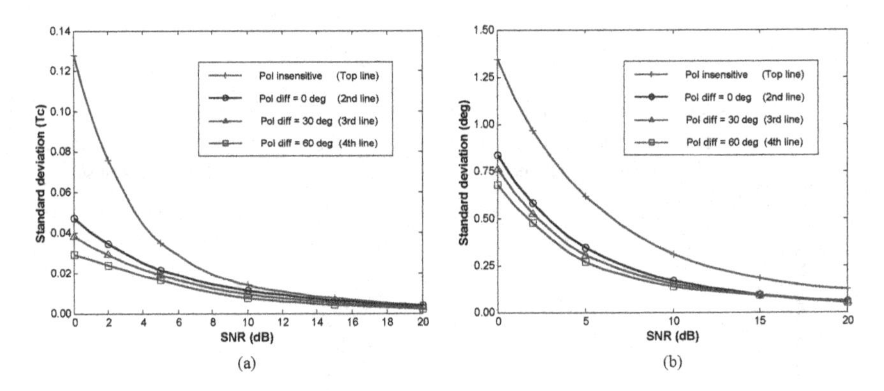

Figure 3.13: Standard deviation of (a) time-of-arrival (TOA) and (b) direction-of-arrival (DOA) estimates for the first multipath of the desired user, in the presence of a second closely-located multipath, versus the SNR.

As expected, paths that are well separated in polarisation exhibit a better standard deviation, with those due to the crossed-dipole array having in general a lower deviation (even when the polarisation difference $\Delta\gamma = 0^\circ$) than that of the polarisation-insensitive array. Such deviation is also observed to be deteriorating with decreasing angular/temporal separation. Take the case whereby user 1 now having two identically polarised ($\gamma = 20^\circ$ and $\eta = 0^\circ$) signal paths, of SNR = 5dB, with its first multipath signal set as $(\theta_{11}, l_{11}) = (80^\circ, 10T_c)$. For a second multipath signal given as $(\theta_{12}, l_{12}) = (80^\circ + \triangle\theta, 11T_c)$, the standard deviation of the estimates associated with the first signal path is plotted against the angular separation $\triangle\theta$ as illustrated in Figure 3.14a. Similarly, Figure 3.14b depicts the standard deviation of the estimates due to the first signal path plotted against the temporal separation $\triangle l$, with its second multipath signal given as $(\theta_{12}, l_{12}) = (81^\circ, 10T_c + \triangle l)$. Again the crossed-dipole array is observed to provide better estimation when compared with that of the polarisation-insensitive array, with both of their estimates degenerating, as anticipated, with decreasing angular and temporal separations.

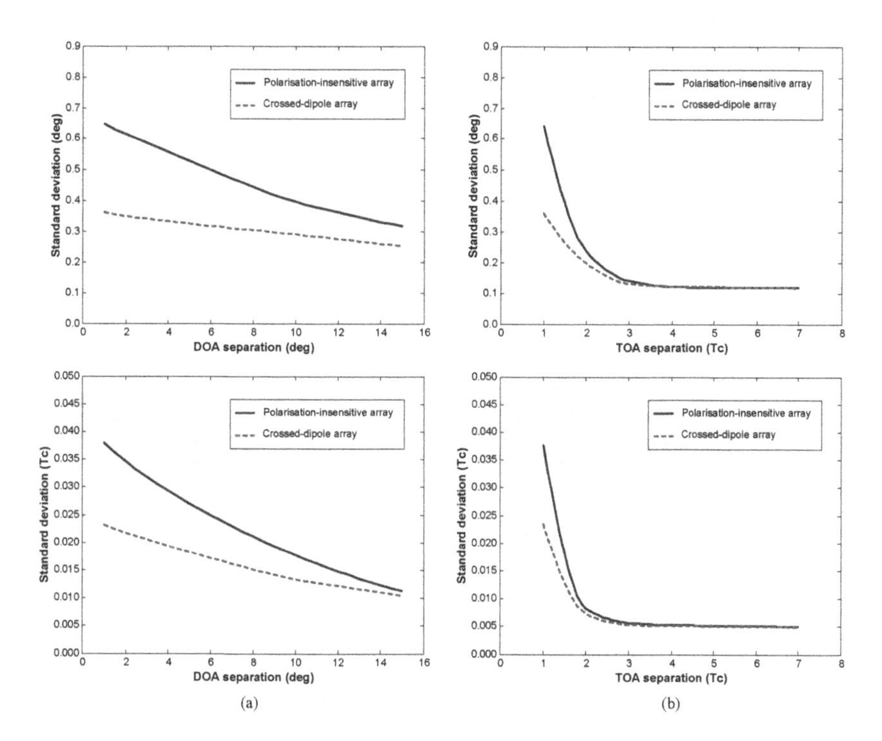

Figure 3.14: Standard deviation of time-of-arrival (TOA) and direction-of-arrival (DOA) estimates for one of the two multipaths of the desired user, vs (a) DOA separation and (b) TOA separation.

Hence it is apparent that through the incorporation of the polarisation diversity in the signal, the crossed-dipole array can improve the accuracy and resolution of the estimates considerably, especially when the signal paths are well separated in polarisation. This improvement is even more significant for paths in a poor SNR environment and paths that are very closely located. To get a clearer picture, let's take a closer look at the effect of polarisation on the spectrum due to the crossed-dipole array as compared with that due to the polarisation-insensitive array. Consider the case of a vertically polarised signal path of $(\theta_{11}, l_{11}, \gamma_{11}, \eta_{11}) = (50°, 5T_c, 90°, 0°)$ together with a closely-located linearly polarised multipath given by $(\theta_{12}, l_{12}, \gamma_{12}, \eta_{12}) = (51°, 6T_c, 30°, 0°)$. Assuming an SNR = -3dB and SIR = -30dB, it is seen from Figure 3.15a that the crossed-dipole array is able to clearly resolve the two signal paths; whereas that of the polarisation-insensitive array has their peaks collapsed together as depicted in Figure 3.15b.

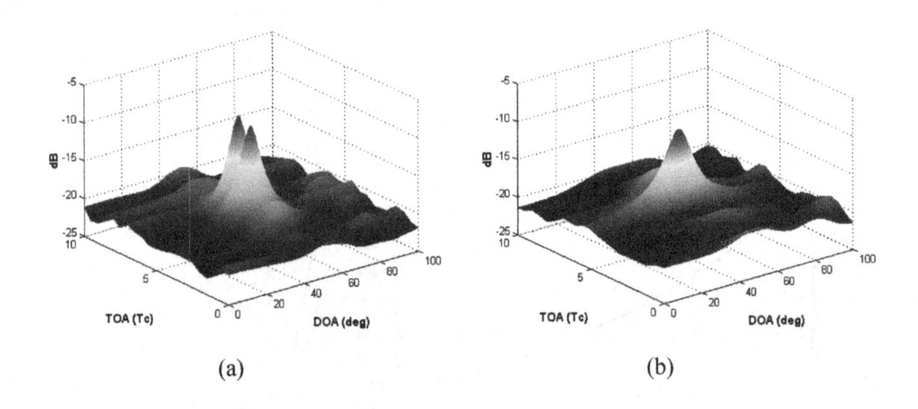

Figure 3.15: Spectrum of two closely-located paths using (a) crossed-dipole array, and (b) polarisation-insensitive array, over an observation interval of 200 data snapshots.

3.5.3 Exemplification of Ill-Conditioned Event

Finally, consider an ill-conditioned situation whereby 2 paths are so closely located that even the use of a crossed-dipole array is unable to distinguish/resolve their spatial and temporal locations. Let's take user 2 as the desired user, with its 4th and 5th paths being so closely located such that the resolution of their peaks collapse and merge into one as illustrated in Figure 3.16a. By making use of the supplementary cost function in Equation (3.36), such an occurrence can be detected as depicted in Figure 3.16b. The capability and proximity limit of such detection is found to be even better if the polarisation difference between the paths widens or the signal strength of the paths increases.

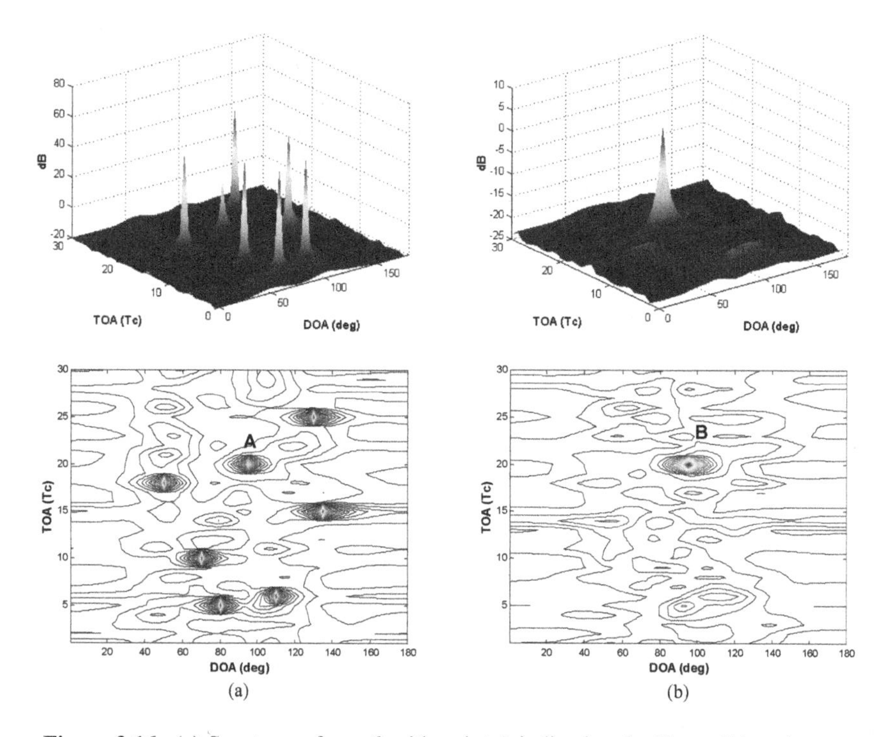

Figure 3.16: (a) Spectrum of user 2 with point A indicating the ill-conditioned event due to the unresolvable 4th and 5th paths, (b) Supplementary spectrum of user 2 with point B indicating the detected ill-conditioned occurrence.

3.6 Summary

An efficient near-far resistant polarisation-angle-delay estimation (PADE) algorithm is presented using a crossed-dipole array for the front-end of the Polar-STAR receiver operating in an asynchronous multipath DS-CDMA system. In contrast to the linearly polarised arrays, the crossed-dipole array is able to capture the polarisation diversity in the signal, hence providing a better detection and estimation/reception performance regardless of its received polarisation. By exploiting the inherent polarisation information in the signal, the joint space-time channel estimator is also able to resolve closely-located paths that otherwise cannot be resolved using its equivalent polarisation-insensitive array. In the event when the paths are so closely located, such that the proposed algorithm fails, its supplementary cost function can be employed to detect such an occurrence. Due to the inclusion of the temporal dimension, the number of multipaths that can be resolved by the algorithm is also no longer constrained by the number of sensors available in the array.

CHAPTER 4

Diffuse-Space-Time Estimation and Reception

In this study, a blind space-time receiver is proposed to handle, under a common theoretical framework, both point and/or diffuse sources for asynchronous multipath DS-CDMA systems. It is known that signal diffusion tends to lead to performance degradation with the application of conventional processing techniques which are based on point sources assumption. A generalised diffusion framework is therefore adopted in the estimation and reception process to cope with such occurrences. The receiver is based on a computationally efficient subspace-type algorithm for its joint space-time channel estimation which is insusceptible to near-far problems. The coherency between the sources is removed by a novel temporal smoothing technique operating in the transformed domain. Unlike many conventional DS-CDMA receivers, the proposed formulation and approach is applicable even with the presence of co-code interferers. Furthermore it is robust to channel estimation errors in the event of any unidentified (incomplete) or erroneous (incorrect) channel parameter.

4.1 Introductory Background

Traditional sensor array processing frequently assumes a simplistic pictorial of the multipath environment which is modelled as an assemblage of point sources. However in a typical wireless communications system, particularly in an urban or suburban setup, the signal transmitted into the channel suffers multiple reflection, refraction, diffraction and scattering effects which will inevitably result in a diffusion of its signal component. This hence leads to a performance degradation when conventional processing techniques, based on point sources scenario, are applied [82]. Thus it is paramount to adopt a generalised diffusion framework in the conception of the estimation and reception process in order to achieve an effective system design.

In the recent years, several experimental measurements [83-86] have been conducted in practical setups to demonstrate the significance and the effect of signal diffusion. For instance, the deteriorating impact of diffuse signals on system performance via beamforming is reported in [83]. These results further emphasise and establish the importance of incorporating the diffusion framework in order to cope with multiple diffuse/point sources environments. But, despite its importance, not many works have been published in the treatment of diffuse sources, especially in the receiver design to handle both diffuse and point sources. Several parametric approaches in the estimation of the diffuse sources have however being introduced. For instance, Meng *et al* suggested the DISPARE (Dispersed Signal Parametric Estimation) [87] method which is based on the weighted projection of the eigenvectors of the signal covariance matrix onto the estimated quasi-noise subspace, with the assumption that the number of sources present at the array are known a priori. Valaee *et al*, on the other hand, developed the DPSE (Distributed Signal Parameter Estimator) [88] algorithm which involves minimising a certain norm of the transformed noise eigenvectors in the signal subspace, provided that the spatial correlation of the distributed signal belongs to a parametric class. In [89], Wu *et al* introduces the vec-MUSIC algorithm by applying the principle of MUSIC based on mathematical operator properties to interpret the geometric structure in the covariance matrix of the vectorised outer-product of the data. All these three techniques: DISPARE, DSPE and vec-MUSIC algorithms, which are basically modifications from classical MUSIC algorithm, provide only the parametric localisation, that is the nominal direction-of-arrival, of the diffuse sources. Other studies taking into account the inclusion of angular spread estimation, in conjunction with its nominal direction-of-arrival estimation, are thenceforth been reported in [90-93]. A computationally-low complexity estimator is proposed in [92] to provide estimates of both the direction-of-arrival and angular spread of the diffuse signal; whereas in [93], the estimates are decoupled and optimised using two successive single-dimensional minimisations. With these two estimates, beamforming techniques, such as the broad-null beamformer devised in [94], can then be applied in its reception process.

In this work, however, an asynchronous blind space-time receiver is proposed, without having to perform angular spread estimation, for both point and/or diffuse multipath DS-CDMA systems [95,96]. Its channel estimation technique is derived from the efficient near-far resistant PADE (polarisation-angle-delay estimation) algorithm, as described in Section 3.4.3, to provide its joint angle and delay estimation. Such adaptation of algorithm from diversely polarised arrays has also been investigated in [97,98] to obtain the direction-of-arrival estimates, but it excludes the temporal dimension in its estimation operation; hence the estimation algorithm in [75] is being attempted in [99], but its estimation procedure is not as compact and efficient and can only be completed after a few tens of observation intervals (bursts). Therefore in this study, a more computationally efficient near-far resistant channel estimation algorithm is being proposed which is to be employed and integrated at the front-end of a novel blind space-time receiver. As a result of its novel underlying architecture, the receiver can also be easily extended to cope in a co-code environment. The resulted receiver is robust against any unidentifiable or erroneous channel parameter resulted in the estimation process. It is optimum with respect to the SIR criterion and is able to achieve (asymptotically) complete interference cancellation. Computer simulation studies demonstrate the effectiveness of the proposed approach and it is shown to perform just as well even when the signal components are co-located in either one or two of the space, time and code domains.

The rest of the chapter is organised as follows. A generalised array manifold vector is first derived in Section 4.2 for both point and/or diffuse sources. Section 4.3 presents the signal mathematical model, based on the generalised array manifold vector, in the point and diffuse DS-CDMA multipath channel. The joint space-time channel estimation algorithm is then described in Section 4.4, together with the formulation of the reception procedure which is easily extendable to a co-code or non co-code environment. Following that, Section 4.5 provides some key simulation studies which depict the performance of the proposed diffusion-based receiver. Finally, the work is concluded in Section 4.6.

4.2 Diffuse Array Manifold Vector

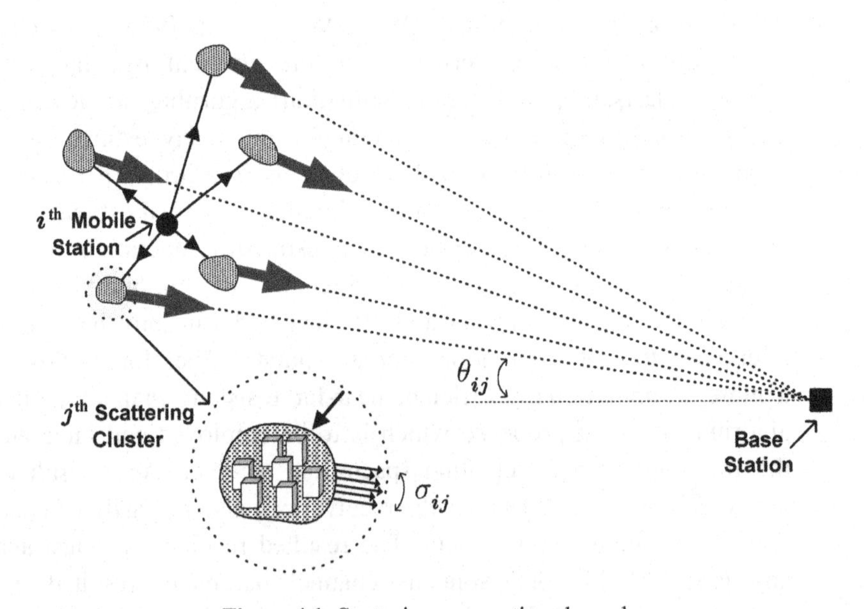

Figure 4.1: Scattering propagation channel.

Consider a transmitted signal due to the ith user arriving at the base station, employing an array of N antenna sensors, via K_i scattering clusters [100] as shown in Figure 4.1. The complex baseband signal vector pertaining to the jth scattering cluster of the ith user, consisting of W_{ij} scatterers, can therefore be modelled as a superposition of all the scatterers' contribution within the cluster, expressed as

$$\underline{\mathrm{s}}_{ij}(t) \quad = \quad \sum_{k=1}^{W_{ij}} \beta_{ijk}\, \underline{S}(\theta_{ij} + \widetilde{\theta}_{ijk})\, m_i(t - \tau_{ijk}) \tag{4.1}$$

where $m_i(t)$ is the ith user's spreaded message signal as described in Equation (2.1), $\underline{S}(\theta)$ is the array response vector pointing towards the azimuth direction θ as delineated in Section 2.3.1, and β_{ijk}, $\widetilde{\theta}_{ijk}$, τ_{ijk} denotes respectively the complex fading coefficient, angular perturbation about the cluster's nominal direction θ_{ij}, and time delay associated with the kth scatterer from the jth cluster of the ith user. The scatterers are grouped together, both spatially and temporally, into clusters as a result of their angular and delay spreads being smaller than the spatial and temporal resolution of the employed system. In particular, the delay differences within each cluster, as compared with its nominal time delay τ_{ij}, are below the system sampling interval T_s, which are thus temporally unresolvable and negligible (i.e. $\tau_{ijk} \approx \tau_{ij}$, for $k = 1, 2, \ldots, W_{ij}$). Additionally, the maximum angular perturbation (i.e. $\sigma_{ij} = 2\max_k\left|\widetilde{\theta}_{ijk}\right|$, where $k = 1, 2, \ldots, W_{ij}$) within the cluster is lower than the system spatial resolution limit such that its contribution to the signal subspace is one-dimensional. Now by taking the first derivative of the array response vector as $\dot{\underline{S}}(\theta) = \partial\underline{S}(\theta)/\partial\theta$, the cluster's baseband signal vector in Equation (4.1) can therefore be remodelled, based on the first-order Taylor approximation, as follows

$$\begin{aligned}
\underline{\mathrm{s}}_{ij}(t) \quad &= \quad \sum_{k=1}^{W_{ij}} \beta_{ijk} \left\{\underline{S}(\theta_{ij}) + \widetilde{\theta}_{ijk}\, \dot{\underline{S}}(\theta_{ij})\right\} m_i(t - \tau_{ij}) \\
&= \quad \kappa_{ij} \underbrace{\left\{\underline{S}(\theta_{ij}) + \varphi_{ij}\, \dot{\underline{S}}(\theta_{ij})\right\}}_{\underline{A}(\theta_{ij},\, \varphi_{ij})} m_i(t - \tau_{ij})
\end{aligned} \tag{4.2}$$

where $\kappa_{ij} = \sum_{k=1}^{W_{ij}} \beta_{ijk}$ and $\varphi_{ij} = \sum_{k=1}^{W_{ij}} \left(\beta_{ijk}\, \widetilde{\theta}_{ijk}\right) \Big/ \sum_{k=1}^{W_{ij}} \beta_{ijk}$ represent the cluster's aggregate fading coefficient and weighted perturbation factor respectively. Notice that in the case of line-of-sight point sources scenario, the above signal diffusion framework in Equation (4.2) can still be utilised since its corresponding angular perturbation will be equated to zero with no angular spreading, i.e. $\varphi_{ij} = 0$.

From the formulation given in Equation (4.2), the array manifold vector due to the jth diffuse cluster of the ith user can be deduced to be

$$\underline{A}_{ij} \triangleq \underline{A}(\theta_{ij}, \varphi_{ij}) = \underline{S}(\theta_{ij}) + \varphi_{ij}\, \underline{\dot{S}}(\theta_{ij}) \tag{4.3}$$

Note that the above diffuse cluster's manifold vector is a function made up of only two parameters, the Direction-of-Arrival θ_{ij} and the weighted perturbation factor φ_{ij}, which consequently simplifies the overall design of the estimation and reception algorithm. And such simplification is made possible by the brought-out of the aggregate fading coefficient κ_{ij} from the diffuse cluster's manifold vector as indicated in Equation (4.2). To appreciate this, let's take for instance the re-expression of the signal vector in Equation (4.2), without the brought-out of the aggregate fading coefficient, as shown below

$$\underline{s}_{ij}(t) = \underbrace{\left\{\kappa_{ij}\, \underline{S}(\theta_{ij}) + \varphi'_{ij}\, \underline{\dot{S}}(\theta_{ij})\right\}}_{\underline{A}(\theta_{ij}, \kappa_{ij}, \varphi'_{ij})} m_i(t - \tau_{ij}) \tag{4.4}$$

where $\kappa_{ij} = \sum_{k=1}^{W_{ij}} \beta_{ijk}$ and $\varphi'_{ij} = \sum_{k=1}^{W_{ij}} \left(\beta_{ijk}\, \widetilde{\theta}_{ijk}\right)$. In this case, the array manifold vector of the diffuse cluster, unlike the one in Equation (4.3), is a function composed of three parameters, that is $\underline{A}(\theta_{ij}, \kappa_{ij}, \varphi'_{ij})$. Such representation, which includes the extra parameter κ_{ij}, will inadvertently impose additional burden on the already multidimensional parametric estimation and reception algorithm. This not only necessitates an extra search dimension requirement in the estimation phase, but also put on an additional unknown uncertainty within the reception process. The simplified expression in Equation (4.3) thus provides a much simpler and less cumbersome manifold representation of the diffuse cluster, without having the consideration of the additional unknown parameter.

4.3 Diffuse-STAR Signal Modelling

The modelling of the received signal in the multiple point and/or diffuse multipath environment can be constructed based on the diffuse cluster's manifold vector defined as $\underline{A}_{ij} \triangleq \underline{A}(\theta_{ij}, \varphi_{ij})$ in Equation (4.3). By denoting $\mathbb{A}_i = [\underline{A}_{i1}, \underline{A}_{i2}, \ldots, \underline{A}_{iK_i}]$, $\underline{\kappa}_i = [\kappa_{i1}, \kappa_{i2}, \ldots, \kappa_{iK_i}]^T$ and $\underline{m}_i(t) = [m_i(t-\tau_{i1}), m_i(t-\tau_{i2}), \ldots, m_i(t-\tau_{iK_i})]^T$, the net continuous baseband vector representation of the received signal, in the presence of additive white Gaussian noise, can therefore be modelled as

$$\underline{x}(t) = \mathbb{A} \, . \, \text{diag}(\underline{\kappa}) \, . \, \underline{m}(t) + \underline{\mathrm{n}}(t) \tag{4.5}$$

where $\underline{\mathrm{n}}(t)$ is the complex Gaussian noise vector and

$$\begin{aligned}
\mathbb{A} &= [\mathbb{A}_1, \ \mathbb{A}_2, \ \ldots, \ \mathbb{A}_M] \\
\underline{\kappa} &= [\underline{\kappa}_1^T, \ \underline{\kappa}_2^T, \ \ldots, \ \underline{\kappa}_M^T]^T \\
\underline{m}(t) &= [\underline{m}_1^T(t), \ \underline{m}_2^T(t), \ \ldots, \ \underline{m}_M^T(t)]^T
\end{aligned}$$

Its discretised version of the received baseband signal vector, on the other hand, can be formulated by extending the diffuse cluster's manifold vector $\underline{A}_{ij}$, via the inclusion of the discretised temporal dimension as demonstrated in Equation (2.16), to form the following Diffuse-Spatial-Temporal ARray (Diffuse-STAR) manifold vector

$$\underline{\mathfrak{h}}_{ij} \triangleq \underline{A}_{ij} \otimes \mathbb{J}^{l_{ij}} \underline{\mathfrak{c}}_i \tag{4.6}$$

Now by taking user 1 as the desired user operating in a M-users co-code environment, in which U_1 of its users share the same code sequence as user 1 (i.e. $\underline{\alpha}_1 = \underline{\alpha}_2 = \cdots = \underline{\alpha}_{U_1}$), the net received discretised diffuse-spatial-temporal signal vector $\underline{x}[n]$ can thus be written as

$$\underline{x}[n] = \bar{\mathbb{H}}_1 \mathbb{G}_1 \underline{\mathrm{a}}_1[n] + \sum_{i=2}^{U_1} \bar{\mathbb{H}}_i \mathbb{G}_i \underline{\mathrm{a}}_i[n] + \sum_{i=U_1+1}^{M} \bar{\mathbb{H}}_i \mathbb{G}_i \underline{\mathrm{a}}_i[n] + \underline{\mathrm{n}}[n] \tag{4.7}$$

where the first term on the right-hand side of Equation (4.7) denotes the discretised signal vector associated with the desired user, the second term contains the interferences due to $(U_1 - 1)$ of its co-code users (arising from either code-reuse interferers or intercepting co-code jammers), the third term represents the remaining interferences from the rest of the users together with their corresponding co-code partners, and the last term is the sampled noise vector. The ith user's data vector constitutes contributions from not only the current but also the previous and next symbols given as $\underline{a}_i[n] = \left[\, a_i[n-1], a_i[n], a_i[n+1]\,\right]^T$, the complex aggregate fading matrix $\mathbb{G}_i = \mathbb{I}_3 \otimes \underline{\kappa}_i$, and the channel matrix $\bar{\mathbb{H}}_i = \left[(\mathbb{I}_N \otimes (\mathbb{J}^T)^{L/2})\mathbb{H}_i,\ \mathbb{H}_i,\ (\mathbb{I}_N \otimes \mathbb{J}^{L/2})\mathbb{H}_i\right]$, with $\mathbb{H}_i$ having columns the Diffuse-STAR manifold vectors corresponding to the ith user, i.e. $\mathbb{H}_i = \left[\underline{\mathfrak{h}}_{i1}, \underline{\mathfrak{h}}_{i2}, \ldots, \underline{\mathfrak{h}}_{iK_i}\right]$.

4.4 Vector Channel Estimation and Reception

4.4.1 Blind Diffuse-Space-Time Estimation

To perform an estimation of the diffuse-space-time channel parameters pertaining to the desired user and all of its co-code interferers, the discretised signal vector $\underline{x}[n]$ in Equation (4.7) is preprocessed by the desired user's preprocessor (i.e. user 1) defined as $\mathbb{Z}_1 = \mathbb{I}_N \otimes \left(\text{diag}(\underline{\tilde{\mathfrak{c}}}_1)^{-1}\mathbb{F}\right)$. Its resultant discretised signal vector, as described in Section 3.4.1, therefore becomes

$$\begin{aligned} \underline{y}[n] &= \mathbb{Z}_1\, \underline{x}[n] \\ &= \sum_{i=1}^{U_1} \tilde{\mathbb{H}}_i\, \underline{\kappa}_i a_i[n] + \mathbb{Z}_1 \underline{I}_{\text{ISI}}[n] + \mathbb{Z}_1 \underline{I}_{\text{MAI}}[n] + \mathbb{Z}_1 \underline{n}[n] \end{aligned} \quad (4.8)$$

where $\tilde{\mathbb{H}}_i = \left[\tilde{\underline{\mathfrak{h}}}_{i1}, \tilde{\underline{\mathfrak{h}}}_{i2}, \ldots, \tilde{\underline{\mathfrak{h}}}_{iK_i}\right]$ with $\tilde{\underline{\mathfrak{h}}}_{ij} = \underline{A}_{ij} \otimes \underline{\Phi}^{l_{ij}}$. Notice that with the application of the preprocessing operation, the signal vector $\underline{y}[n]$ is effectively decoupled into four terms, with its first term containing the desired signal constituent $\tilde{\mathbb{H}}_1\, \underline{\kappa}_1 a_1[n]$ together with its $(U_1 - 1)$ co-code partners, followed by the ISI, MAI and noise components respectively.

Similarly the second order statistics of the above preprocessed signal vector would bring about a collapse in the desired signal subspace dimension down to unity. The temporal smoothing technique thus has to be performed to restore the rank of this subspace, now consisting in addition all other users sharing the same desired user's code sequence, back to a dimension of $\sum_{i=1}^{U_1} K_i$. However this time, instead of carrying out the smoothing procedure directly on the signal vector, which involves multiple subvector covariance calculations, a more efficient procedure will be devised based on the preprocessed signal vector covariance matrix $\mathbb{R}_{\underline{y}} \in \mathcal{C}^{NL \times NL}$.

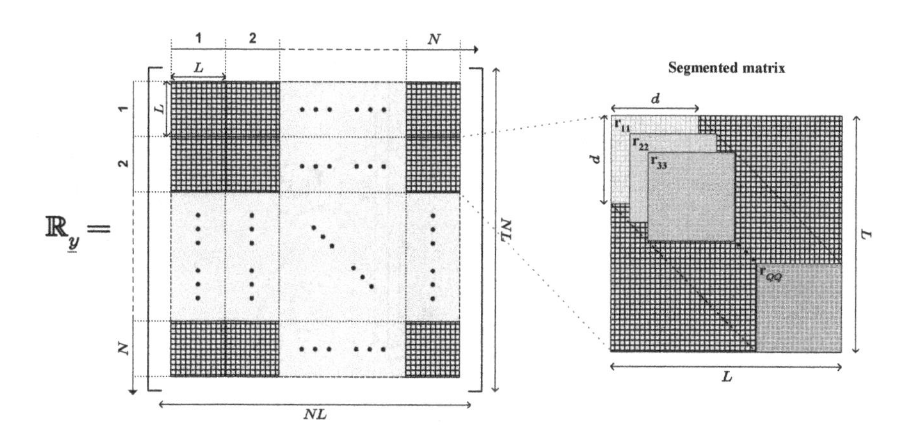

Figure 4.2: Temporal smoothing technique.

First let's partition the matrix $\mathbb{R}_{\underline{y}}$ into N^2 segmented matrices of size $L \times L$ as illustrated in Figure 4.2. By averaging a set of Q (where $Q = L - d + 1$ and $d < L$) overlapping $d \times d$ submatrices along the principal diagonal of each of the segmented matrices simultaneously, a $Nd \times Nd$ temporal-smoothed covariance matrix $\mathbb{R}_{\text{Tsmooth}}$ is therefore constructed. However its dimensionality can only be successfully restored unless $Q \geq \sum_{i=1}^{U_1} K_i$. Note that the smoothing technique is applicable even in situation with multiple clusters arriving from the same direction (co-directional). But for multiple clusters of a particular user arriving at the same time (co-delay), spatial smoothing technique, developed for a uniform linear array [101], has to be overlaid on top of $\mathbb{R}_{\text{Tsmooth}}$ to form the spatial-temporal-smoothed covariance matrix $\mathbb{R}_{\text{STsmooth}}$. This can be done by subdividing and grouping every $d \times d$ elements within the matrix $\mathbb{R}_{\text{Tsmooth}}$ as $\mathbb{R}_{xy}$ (for $x, y = 1, 2, \ldots, N$) as depicted in Figure 4.3. Next, P overlapping submatrices each of size $ed \times ed$ (where $e = N - P + 1$) are then extracted from the matrix's main diagonal. These submatrices, which are essentially the subarray covariance matrices, are finally averaged together to form the smaller covariance matrix $\mathbb{R}_{\text{STsmooth}}$.

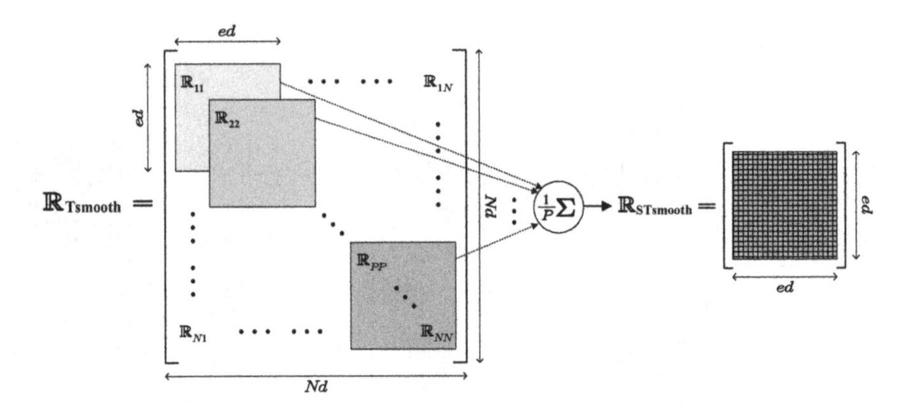

Figure 4.3: Spatial smoothing technique.

Having obtained the smoothed covariance matrix, the following channel estimation cost function pertaining to the clusters can be devised based on the preprocessed Diffuse-STAR manifold vector $\underline{\tilde{\mathfrak{h}}}$, given as

$$\xi(\theta, l, \varphi) = \underline{\tilde{\mathfrak{h}}}^H \, \mathbb{E}_{\text{n}} \mathbb{E}_{\text{n}}^H \, \underline{\tilde{\mathfrak{h}}} \tag{4.9}$$

where $\underline{\tilde{\mathfrak{h}}} = \underline{A} \otimes \underline{\Phi}_d^l$ in which $\underline{\Phi}_d$ is a subvector of $\underline{\Phi}$ with length d, and $\mathbb{E}_{\text{n}}$ has columns the generalised noise eigenvectors of $(\mathbb{R}_{\text{STsmooth}}, \mathbb{D})$ with $\mathbb{D}$ representing the spatial-temporal-smoothed diagonal matrix $\mathbb{Z}_1 \mathbb{Z}_1^H$. Notice that for a given environment consisting of solely point multipath sources, the array manifold vector as stated in Equation (4.3) becomes

$\underline{A}_{ij} = \underline{S}_{ij}$ since $\varphi_{ij} = 0$ for $i = \{1, 2, \ldots, M\}$ and $j = \{1, 2, \ldots, K_i\}$. This consequently leads to a rather simple two-dimensional space-time cost function given as $\xi(\theta, l) = (\underline{S} \otimes \underline{\Phi}_d^l)^H \mathbb{E}_n \mathbb{E}_n^H (\underline{S} \otimes \underline{\Phi}_d^l)$. However for the case as shown in Equation (4.9), its minimisation operation requires a multidimensional search over the three parameters θ, l and φ. In order to reduce the size of this search space, the manifold vector $\underline{A}$ employed within the preprocessed Diffuse-STAR manifold vector $\underline{\tilde{\mathfrak{h}}}$ can be rewritten alternatively as $\underline{A}(\theta, \varphi) = \mathbb{M}(\theta) \,.\, \underline{p}(\varphi)$ where

$$\mathbb{M}(\theta) = \begin{bmatrix} \underline{S}(\theta) & \underline{\dot{S}}(\theta) \end{bmatrix} \quad ; \quad \underline{p}(\varphi) = [1 \quad \varphi]^T \tag{4.10}$$

But note that for a linear array geometry, for instance an array lying along the x-axis, the term $\underline{\dot{S}}(\theta)$ will be nullified for $\theta = 0$ and π due to its derivative coefficient being proportional to an azimuth-dependent function $\sin(\theta)$. In this case, the true $\underline{\dot{S}}(\theta)$ can be replaced by lumping its azimuth-dependent function together with φ, i.e. $\mathbb{M}(\theta) = [\underline{S}(\theta),\ \underline{D}(\theta)]$ and $\underline{p}(\varphi, \theta) = [1,\ \varphi\sin(\theta)]^T$ where $\underline{D}(\theta)$ is the derivative term $\underline{\dot{S}}(\theta)$ after discarding its azimuth-dependent function.

The re-expression in Equation (4.10) thus yields a cost function which is similar to that commonly-used for diversely polarised array, i.e.

$$\xi(\theta, l, \varphi) = \underline{p}^H \underbrace{(\mathbb{M} \otimes \underline{\Phi}_d^l)^H \mathbb{E}_n \mathbb{E}_n^H (\mathbb{M} \otimes \underline{\Phi}_d^l)}_{\mathbb{V} \triangleq \begin{bmatrix} v_{11} & v_{12} \\ v_{21} & v_{22} \end{bmatrix}} \underline{p} \tag{4.11}$$

As with the diversely polarised arrays scenario, the search over the two-dimensional $\underline{p}$ space is equivalent to finding the eigenvectors associated with the smaller eigenvalues of $\mathbb{V}$. Thus by establishing the minimisation operation based on the smaller eigenvalue of the 2×2 matrix $\mathbb{V}$, the Diffuse-STAR cost function in Equation (4.11) can hence be reduced to

$$\xi(\theta, l) = \text{trace}(\mathbb{V}) - \sqrt{\text{trace}(\mathbb{V})^2 - 4\det(\mathbb{V})} \tag{4.12}$$

which provides the joint estimation of the cluster's nominal Direction-of-Arrival (DOA) and Time-of-Arrival (TOA). Its corresponding weighted perturbation factor, on the other hand, is estimated as follows

$$\widehat{\varphi} = \begin{cases} (\xi_{\min} - 2v_{11})/2v_{12} \\ \text{or} \\ 2v_{21}/(\xi_{\min} - 2v_{22}) \end{cases} \tag{4.13}$$

where $\xi_{\min}$ is the path's minima value using the Diffuse-STAR cost function in Equation (4.12). However in the case of a linear array, the estimated weighted perturbation factor $\widehat{\varphi}$ has to be divided by its azimuth-dependent term $\sin(\widehat{\theta})$. Note that further simplification can be attained in the above expression since the value $\xi_{\min}$ is normally trivial.

4.4.2 Co-code/Non Co-code Diffusion-Based Reception

Now for a set of B unique PN-code signatures, the estimated composite channel parameters can be obtained as

$$\widehat{\mathbb{H}}_{com} = \left[\widehat{\mathbb{H}}_{PN,1}, \quad \widehat{\mathbb{H}}_{PN,2}, \quad \ldots, \quad \widehat{\mathbb{H}}_{PN,B}\right] \tag{4.14}$$

where $\widehat{\mathbb{H}}_{PN,u}$ has columns the estimated Diffuse-STAR manifold vectors associated with all the users sharing the uth PN-code signature. In order for the blind space-time receiver to suppress the contributions from the MAI and ISI interferences, it is proposed that the received discretised signal vector $\underline{x}[n]$ in Equation (4.7) be passed through a novel multi-cluster filter bank given as

$$\underline{z}[n] = \mathbb{L}^H . \underline{x}[n] \tag{4.15}$$

where $\mathbb{L}$ is the multi-cluster filter bank based on the orthogonal projection of the interference subspace, i.e.

$$\mathbb{L} = \mathbb{P}^{\perp}_{\widehat{\mathbb{H}}_{intf}} \widehat{\mathbb{H}}_{PN,1} \left(\widehat{\mathbb{H}}^H_{PN,1} \mathbb{P}^{\perp}_{\widehat{\mathbb{H}}_{intf}} \widehat{\mathbb{H}}_{PN,1}\right)^{-1} \tag{4.16}$$

where $\mathbb{P}^{\perp}_{\mathbb{K}} = \mathbb{I} - \mathbb{K}(\mathbb{K}^H \mathbb{K})^{-1}\mathbb{K}^H$ is the complementary projection operator of the matrix $\mathbb{K}$; and $\widehat{\mathbb{H}}_{intf} = \left[(\mathbb{I}_N \otimes (\mathbb{J}^T)^{L/2})\widehat{\mathbb{H}}_{com}, \widehat{\mathbb{H}}^{(PN,1)}_{com},\right.$ $\left.(\mathbb{I}_N \otimes \mathbb{J}^{L/2})\widehat{\mathbb{H}}_{com}\right]$, in which $\widehat{\mathbb{H}}^{(PN,1)}_{com}$ is the estimated composite channel parameters $\widehat{\mathbb{H}}_{com}$ with the exclusion of the matrix $\widehat{\mathbb{H}}_{PN,1} = \left[\widehat{\mathbb{H}}_1, \widehat{\mathbb{H}}_2,\right.$ $\left.\ldots, \widehat{\mathbb{H}}_{U_1}\right]$ corresponding to the desired user's PN-code sequence.

In a non co-code environment, the outputs of the filter bank can hence be simply combined to realise the decision statistic for the nth symbol of the desired user

$$b[n] \quad = \quad \underline{w}^H . \underline{z}[n] \tag{4.17}$$

where $\underline{w}$ is the combining weight vector obtained using the principal eigenvector of the autocorrelation matrix of Equation (4.15). But such methodology will result in an arbitrary phase ambiguity, which can however be easily resolved by differential encoding with the transmitted data symbol $e_i[n] = e_i[n-1].\mathrm{a}_i[n]$ and the decoding according to the following criterion $\widehat{e}_i[n] = \mathrm{sign}(Re\{b[n-1]^*.b[n]\})$.

On the other hand, in the presence of co-code users, the filter bank outputs, comprising of $\sum_{i=1}^{U_1} K_i$ branches, need to be partitioned to differentiate those branches belonging to the desired user. This can be done by assigning a short random header sequence common to the designated non co-code user group. The filter bank output, following the decision device, are then passed to the *Branch Identification Process*, whereby a correlation with the header is performed to identify the dominant branch exhibiting the strongest presence of the header sequence. The dominant branch is subsequently cross-correlated with each of the $\sum_{i=1}^{U_1} K_i$ branches and compared with a prespecified threshold value. When the correlated output exceeds the threshold value, it is assigned to the desired user. Having segregated the branches, the desired user's filter bank outputs can then be combined using Equation (4.17). Note that it is not necessary to assign all the K_1 branches to the desired user. If the channel parameters of any particular branch is erroneous (incorrect channel estimation) or unidentified (incomplete channel estimation), the *Branch Identification Process* will leave that branch unassigned, thus inducing robustness to the receiver. In a similar manner, the above technique can also be applied to decode the remaining non co-code users in the designated group as well as their corresponding co-code partners.

The main steps of the proposed Diffuse-STAR receiver incorporating the diffusion framework to handle both point and/or diffuse sources in a co-code or non co-code environment is outlined as shown below:

1. Sample the array output and concatenate the tapped-delay lines (TDL) contents to form the discretised signal vector $\underline{x}[n]$.

2. Apply the desired user's preprocessor onto the discretised signal vector $\underline{x}[n]$ as shown in Equation (4.8). Form the matrix $\mathbb{R}_{\mathrm{STsmooth}}$ by performing the spatial-temporal smoothing technique to restore the dimensionality of the desired signal subspace.

3. Carry out the generalised eigenvector decomposition of $\mathbb{R}_{\mathrm{STsmooth}}$ and apply the cost function in Equation (4.12) to obtain the joint space-time channel estimation of the point and/or diffuse signals. Its corresponding weighted perturbation factor estimation, on the other hand, is obtained using Equation (4.13).

4. Pass the discretised signal vector through the multi-cluster filter bank which is computed based on the estimated channel parameters as described in Equations (4.15) and (4.16).

   - If the multiuser environment is non co-code, simply combine the outputs of the filter bank, as shown in Equation (4.17), by using the principal eigenvector of the autocorrelation matrix of Equation (4.15).

   - If the environment is co-code, apply the *Branch Identification Process* to segregate those output branches belonging to the desired user and combine them using Equation (4.17).

4.5 Simulation Studies

To demonstrate the key features of the proposed blind space-time receiver, let's consider a uniform $N = 5$ element circular array (with half-wavelength spacing) operating in the presence of $M = 7$ co-channel DS-CDMA users, each being assigned a Gold sequence of length $\mathcal{N}_c = 31$ (of rectangular chip pulse-shaping). The number of unique code signatures is $B = 3$, with the designated user group of interest consisting of $i = 1$, 3 and 6, as listed in Table 4.1. The angular spread σ (SPD) of each cluster is rendered by generating 40 spatially uniform-distributed local scatterers. And the array is assumed to chip-rate sample 150 data symbols for processing.

Table 4.1: Users' parameters

Code signature vector 1 ($\underline{\alpha}_1$)				Code signature vector 2 ($\underline{\alpha}_2$)				Code signature vector 3 ($\underline{\alpha}_3$)			
(user i, cluster j)	DOA	TOA	SPD	(user i, cluster j)	DOA	TOA	SPD	(user i, cluster j)	DOA	TOA	SPD
$(i, j) = (1, 1)$	120°	$5T_c$	0°	$(i, j) = (3, 1)$	230°	$4T_c$	1.9°	$(i, j) = (6, 1)$	160°	$14T_c$	2.6°
$(i, j) = (1, 2)$	170°	$25T_c$	1.3°	$(i, j) = (3, 2)$	250°	$10T_c$	3.4°	$(i, j) = (6, 2)$	190°	$30T_c$	4.5°
$(i, j) = (1, 3)$	250°	$10T_c$	0.5°	$(i, j) = (3, 3)$	350°	$22T_c$	5.0°	$(i, j) = (6, 3)$	200°	$28T_c$	1.7°
$(i, j) = (1, 4)$	300°	$20T_c$	4.2°	$(i, j) = (4, 1)$	50°	$4T_c$	4.2°	$(i, j) = (6, 4)$	200°	$21T_c$	0°
$(i, j) = (2, 1)$	60°	$10T_c$	2.4°	$(i, j) = (5, 1)$	80°	$18T_c$	0°	$(i, j) = (6, 5)$	320°	$16T_c$	3.8°
$(i, j) = (2, 2)$	170°	$15T_c$	0°	$(i, j) = (5, 2)$	120°	$15T_c$	0.1°	$(i, j) = (7, 1)$	10°	$3T_c$	0.9°

4.5.1 Performance of Diffuse-Space-Time Algorithm

Let's assume that user 1 is the desired user having an input SNR of 20dB; while its corresponding co-code partner (i.e. user 2) constitutes an interference ratio of 0dB, together with the rest of the remaining interferers each having a SIR = -20dB (i.e. near-far problem). By setting $d = 55$ (that is $Q = 8$) for temporal smoothing, it is seen from Figure 4.4 that all the clusters, be it diffuse or line-of-sight occurrences, associated with the desired user as well as its co-code partners can be identified/estimated successfully using the proposed algorithm. Notice that the algorithm can still operate even in situations whereby (i) the 2nd

cluster of user 1 is co-located in both the code and space domains with the 2nd cluster of user 2, (ii) the 3rd cluster of user 1 is co-located in both the code and time domains with the 1st cluster of user 2, and (iii) the 3rd cluster of user 1 is co-located in both the space and time domains with the 2nd cluster of user 3. In addition to that, the number of clusters that can be resolved by the algorithm is also not constrained by the number of sensors available in the array.

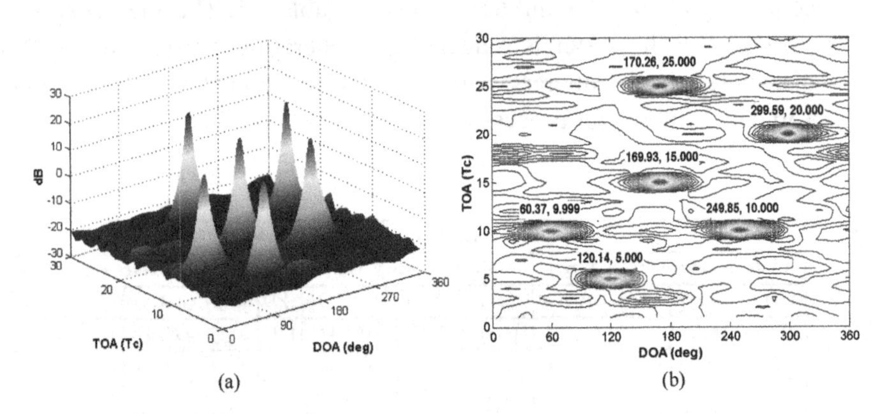

Figure 4.4: MUSIC-type spectrum of the desired user (i.e. user 1 with code signature vector $\underline{\alpha}_1$) and its co-code partners.

The clusters belonging to the desired user are then singled out from its corresponding co-code partner, as illustrated in Table 4.2, by applying the *Branch Identification Process* with a header sequence of length 7. Figure 4.5 depicts the signal constellation of (a) the proposed diffusion-based receiver (i.e. adhering to $\underline{A}(\theta,\varphi) = \underline{S}(\theta) + \varphi\dot{\underline{S}}(\theta)$ framework) as compared with (b) its equivalent non-diffusion counterpart (i.e. adhering to $\underline{A}(\theta) = \underline{S}(\theta)$ framework), in conjunction with (c) a diffusion-based ST (Space-Time) decorrelating detector, and (d) a diffusion-based 2D (Two-Dimensional) RAKE receiver, with the latter three assuming full knowledge of the channel information (including knowledge of all the desired user's co-code interferences). It is thus clear that by incorporating the diffusion framework, the proposed diffusion-based receiver offers a considerable performance enhancement than that of its non-diffusion counterpart. Furthermore, even with the adoption of a common diffusion framework for comparison, the proposed receiver is observed to perform significantly much better than that of the 2D RAKE receiver, with its performance approaches that of the ST decorrelating detector.

Table 4.2: Branch Identification Process (User 1)

(i^{th} user, j^{th} cluster)	(1, 1)	(1, 2)	(1, 3)	(1, 4)	(2, 1)	(2, 2)
Correlated output	1.0000	0.9999	1.0000	1.0000	0.0151	0.0151

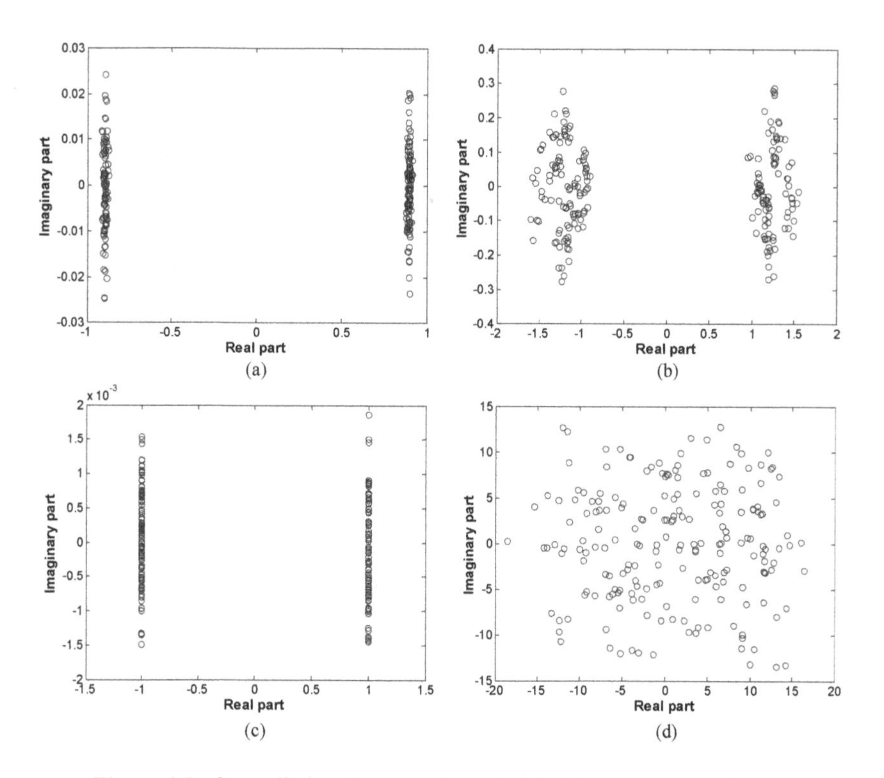

Figure 4.5: Constellation diagram of user 1 with (a) the proposed diffusion-based receiver, (b) its equivalent non-diffusion counterpart, (c) the diffusion-based ST decorrelating detector and (d) the diffusion-based 2D RAKE receiver.

In a similar manner, the constellation diagram of its co-code partner, that is user 2, can also be obtained as illustrated in Figure 4.6a. Thus far, the near-far resistant capability of the proposed algorithm is examined in the presence of strong non co-code interferers, with each contributing a SIR = -20dB. Now, let's consider the effect of having a near-far situation contributed by its corresponding co-code partner. Using the same scenario as above, the power of user 1 is increased 30dB above that of its co-code partner, user 2. By applying the same diffuse-space-time algorithm, the resulted constellation diagram of user 2 is plotted as depicted in Figure 4.6b. Comparing with the strong non co-code interferers case as seen in Figure 4.6a, little deteriorating effect is observed from the resulted signal constellation. It is therefore conclusive that the algorithm is near-far resistant regardless of whether the interferers come from co-code or unique-code groups.

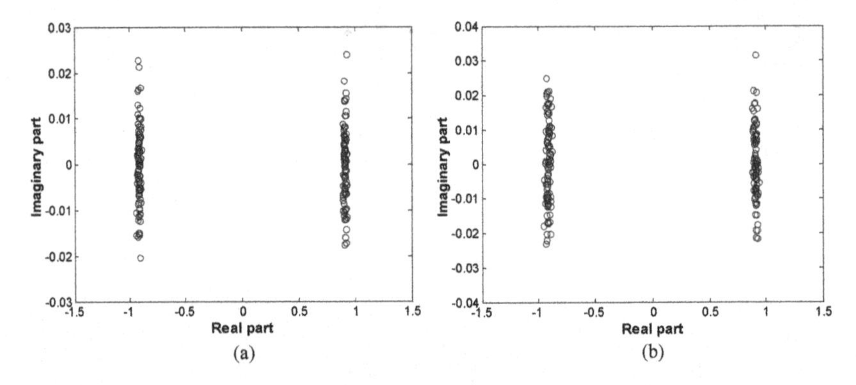

Figure 4.6: Constellation diagram of co-code partner (user 2) with user 1 constituting (a) an interference ratio of 0dB and (b) an interference ratio of 30dB.

4.5.2 Robustness to Channel Estimation Errors

To illustrate the robustness of the receiver against channel estimation errors, let's take user 3 as the desired user in the designated group, with its 1st cluster yielding erroneous channel parameters and its 3rd cluster unidentified in the channel estimation process. Its corresponding space-time spectrum is plotted as shown in Figure 4.7. Note that its 1st cluster $(230°, 4T_c)$ yields an erroneous channel parameters of $(179.87°, 6.999T_c)$ and its 3rd cluster $(350°, 22T_c)$ is unidentified as can be seen from the 2D spectrum plot of Figure 4.7b.

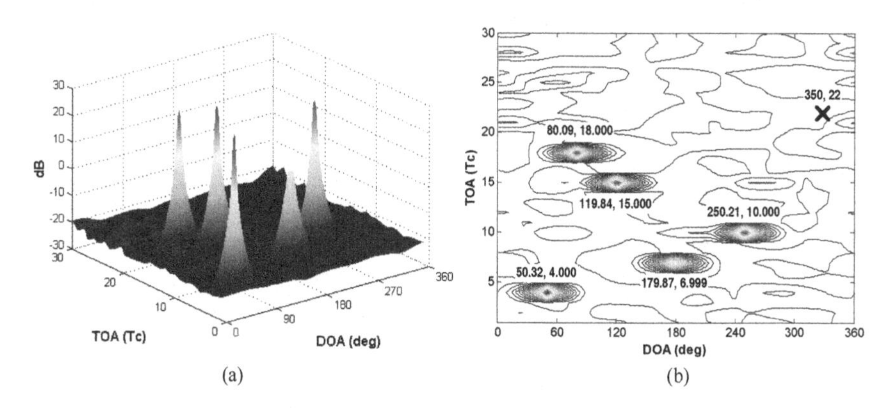

Figure 4.7: Space-time spectrum of user 3 together with its corresponding co-code partners (all sharing the code signature vector $\underline{\alpha}_2$).

Similarly, by applying the *Branch Identification Process* (as shown in Table 4.3) and setting the prespecified threshold value to, say 0.5, the constellation diagram of user 3 employing (a) the proposed receiver is compared with (b) its corresponding diffusion-based ST decorrelating detector as depicted in Figure 4.8. As a result of the robustness induced by the *Branch Identification Process* in the receiver architecture, the signal constellation due to the proposed receiver is seen to be much better and more well-defined than that of the ST decorrelating detector. Notice that the only identifiable and estimable cluster produced by the *Branch Identification Process*, that is the 2nd cluster of user 3, is co-located both in the space and time domains with the 3rd cluster of user 1. It is therefore evident that as long as there exists a single identifiable and estimable cluster due to the desired user separable in at least one of the space, time or code domains, the proposed receiver will be robust to any erroneous (incorrect channel estimation) or unidentified (incomplete channel estimation) channel parameters incurring at the front-end estimation phase.

Table 4.3: Branch Identification Process (User 3)

(i^{th} user, j^{th} cluster)	(3, 1)	(3, 2)	(3, 3)	(4, 1)	(5, 1)	(5, 2)
Correlated output	0.1960	1.0000	n.a.	-0.0754	0.0352	0.0352

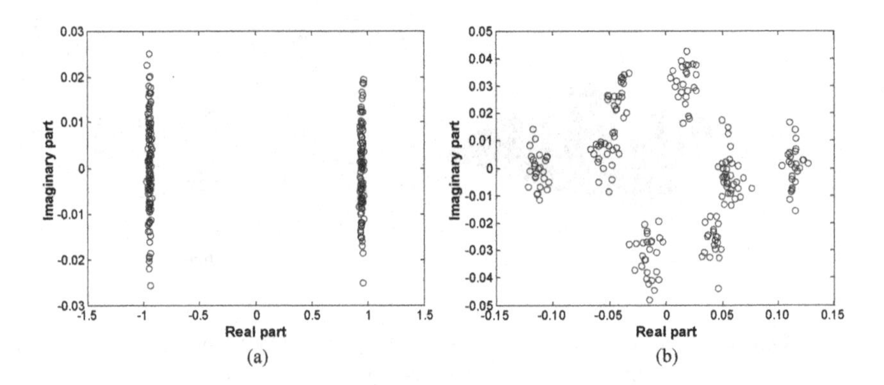

Figure 4.8: Constellation diagram of user 3 with diffusion-based (a) proposed receiver and (b) ST decorrelating detector, in the presence of incorrect or incomplete channel estimation.

4.5.3 Evaluation of Diffusion Framework Reception

Next, let's take a closer look at the performance of the receiver with framework based on (i) the proposed diffusion model $\underline{A}(\theta, \varphi) = \underline{S}(\theta) + \varphi \dot{\underline{S}}(\theta)$ as compared with that based on (ii) the conventional non-diffusion model $\underline{A}(\theta) = \underline{S}(\theta)$. For comparison clarity, let's assume that the all the clusters under consideration have the same angular spread σ. Using the same setting as in Table 4.1, the output SNIR of the receivers are averaged over 1000 Monte Carlo runs and plotted as shown in Figure 4.9. It is therefore apparent that by incorporating the diffusion aspect into the framework, the proposed receiver provides a substantial performance gain than that based on the conventional model which is observed to be deteriorating with the angular spread σ.

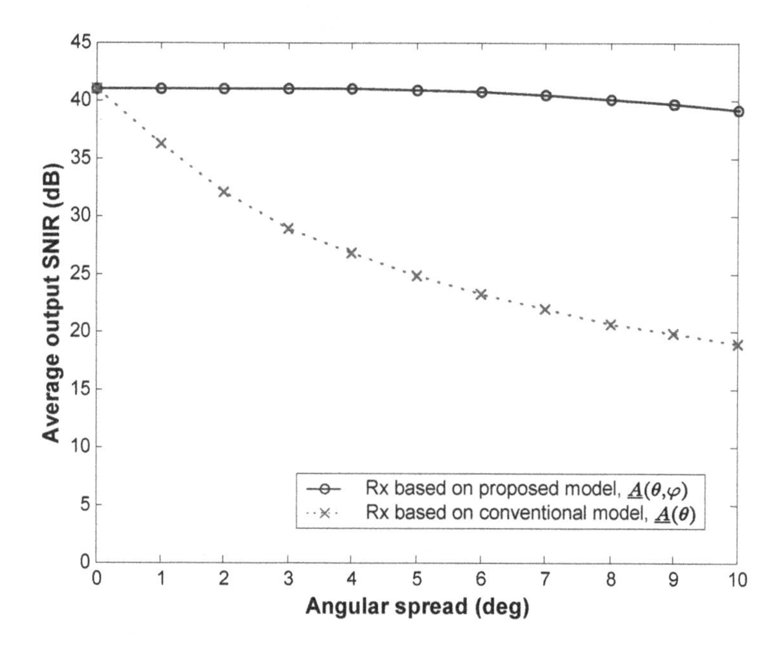

Figure 4.9: SNIR performance of receiver with framework based on (i) proposed model $\underline{A}(\theta, \varphi) = \underline{S}(\theta) + \varphi \dot{\underline{S}}(\theta)$, and (ii) conventional model $\underline{A}(\theta) = \underline{S}(\theta)$, versus the angular spread σ.

Having seen the benefit of including the diffusion model in the receiver framework, let's assess the performance of the proposed receiver architecture in terms of output SNIR versus input SNR, together with the conventional ST decorrelating detector and 2D RAKE receiver. For fairness in comparison, the diffusion framework will be integrated into all the receivers under consideration. As expected, the output SNIR of the proposed receiver, as illustrated in Figure 4.10, is much higher than that of the RAKE receiver; and its performance closely follows that of the ST decorrelating detector.

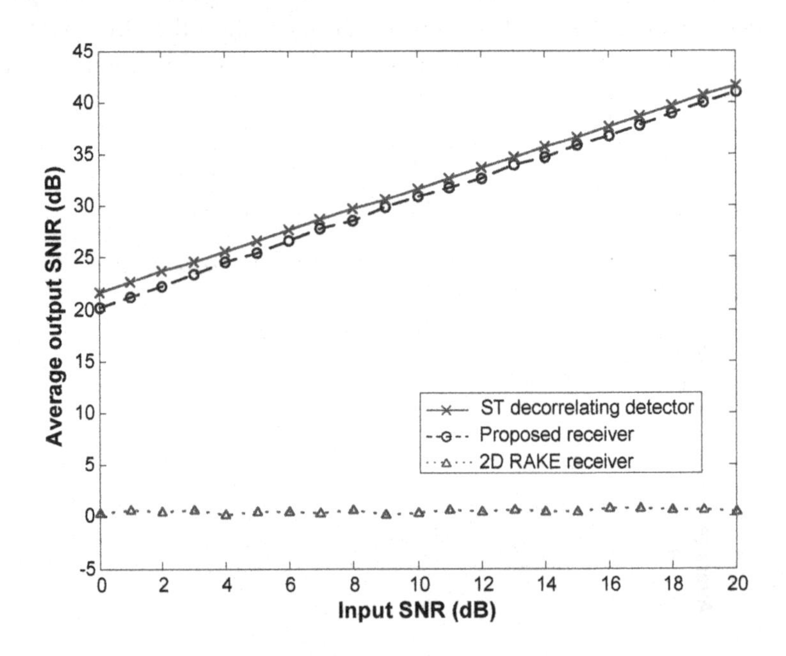

Figure 4.10: Output SNIR versus input SNR of (i) proposed receiver, (ii) ST decorrelating detector, and (iii) 2D RAKE receiver.

4.6 Summary

A blind space-time receiver based on a computationally efficient near-far resistant channel estimation technique is presented for multiple point and/or diffuse signal paths in an asynchronous DS-CDMA system. Due to its underlying architecture, the receiver can be easily extended to cope in a non-unique code signature environment by simply adding a novel *Branch Identification Process* in the detection procedure. The receiver is robust to erroneous or incomplete channel estimation since its operation requires only the existence of an estimable cluster due to the desired user separable/identifiable in at least one of the following domains: space, time and code. The number of resolvable clusters is also no longer limited by the number of sensors available in the array.

CHAPTER 5

Doppler-Space-Time Estimation and Reception

A blind near-far resistant MIMO array receiver is proposed, in this study, for time-varying asynchronous multipath DS-CDMA systems. Unlike many other MIMO works, the proposed receiver is built on *array-antenna* technology which is commonly used in applications such as the well-known smart-antenna system. As such, a novel Doppler-space-time estimation algorithm can be devised which provides the joint angle and delay, as well as the Doppler shift frequency estimation. Its multipath's Doppler effect, instead of being detrimental to the system performance, is been exploited to provide an extra domain of diversity. The proposed MIMO array receiver requires no knowledge of the channel and is insusceptible to near-far problems. It is applicable in either of the two MIMO transmission schemes: (i) data rate maximisation MIMO scheme or (ii) diversity maximisation MIMO scheme. Furthermore, the spreading code assignment strategies employed in typical CDMA BLAST system can also be adapted in the following work.

5.1 Introductory Background

Telecommunications have witnessed an explosive growth in mobile devices which consequently create a huge market opportunity for wireless data access. In particular, broadband wireless access along with evolving mobile internet and multimedia services are driving the recent surge of research and development activities to meet the demanding requirements of providing very high data rates with ubiquity, mobility, and portable characteristic in a cellular system. But the constraint in power at each mobile terminal calls for a more reliable detection/ reception technique to be employed in such cellular system. The increased use of antenna elements at both ends of the transmission link, giving rise to the research in Multiple Input and Multiple Output (MIMO) systems [102], is one of the viable and promising means of enabling the types of data rates and capacities needed for wireless internet and multimedia services.

However a vast majority of the reported works on MIMO systems often adopt the assumption of multiple independent antenna elements [103-106], referred herein as *multiple-antenna* system. This is different from an *array-antenna* system where a number of antenna elements form an array system of a given geometry with measurements taken with respect to the array reference point. The powerful effect of *array-antenna* system, such as smart-antennas lies in beamforming whereby its mainlobe is focused in particular desired directions [107-109]. Such array processing and communications techniques, exploiting the structure of *array-antenna* technology, have evolved into a well-established technology [110], moving from old conventional direction nulling and phase-arrays to advanced superresolution arrays, for instance the polarised superresolution arrays in [74-76]. By harnessing the spatial-temporal properties of the channel provided by the *array-antenna*, an extra layer of co-channel interference cancellation and new ways for handling unwanted channel effects, such as Doppler spread and fading, can be developed [111].

Unlike most MIMO researches which require the full knowledge of the channel, this work proposes a blind near-far resistant MIMO array receiver for time-varying asynchronous multipath DS-CDMA system

[112]. The proposed receiver is applicable in both the data rate or diversity maximisation transmission schemes. Although a single spreading code is being assigned to each user in the work here, the proposed formulation can also be easily adapted to the code assignment strategies employed in typical spread spectrum BLAST system. Through the incorporation of the *array-antenna* system, a novel subspace-based Doppler-space-time estimation algorithm is devised which provides the joint angle and delay, as well as the Doppler shift frequency estimation. The multipath Doppler spread, which is often regarded as one of the detrimental factors in degrading the performance of existing receivers, is being employed in the proposed receiver to provide an additional form of diversity [113,114]. In addition to that, the structural property of the Doppler component is put to good use in the development of the joint space-time estimation and Doppler frequency estimation algorithms. The resulting receiver, with its front-end integrated with the channel estimator, is also robust against any unidentifiable or erroneous channel parameter resulted in the estimation process. It is optimum with respect to the SIR criterion and is able to achieve (asymptotically) complete interference cancellation.

The structure of the remaining sections is as follows. In Section 5.2, the spatial array manifold vector associated with the MIMO *array-antenna* system is first introduced, and later modified to include the temporal dimension to form the Spatial-Temporal ARray (STAR) manifold vector. The Doppler component is then incorporated to construct the Doppler-STAR manifold vector, which is to be used later in the modelling of the MIMO array system operating in a time-varying asynchronous multipath DS-CDMA environment. Having obtained that, a computationally efficient near-far resistant joint space-time channel estimation, in conjunction with its Doppler estimation, algorithm is proposed in Section 5.3, together with the formulation of a novel blind robust MIMO receiver design. Following that, Section 5.4 provides several simulation studies which depict the performance of the blind MIMO array receiver, highlighting the key potential benefits of utilising the *array-antenna* system. The work is then concluded in Section 5.5. For completeness, recommendation in the use of the algorithm is also included at the end of the study in Section 5.6.

5.2 MIMO Array System Formulation

5.2.1 Continuous Time-Varying Signal Model

Consider an M-users asynchronous DS-CDMA multiple input multiple output system, with each user having N_t transmitting antenna elements. Two transmission schemes, as shown in Figure 5.1, can be employed over the MIMO channels: (i) data rate maximisation scheme or (ii) diversity maximisation scheme.

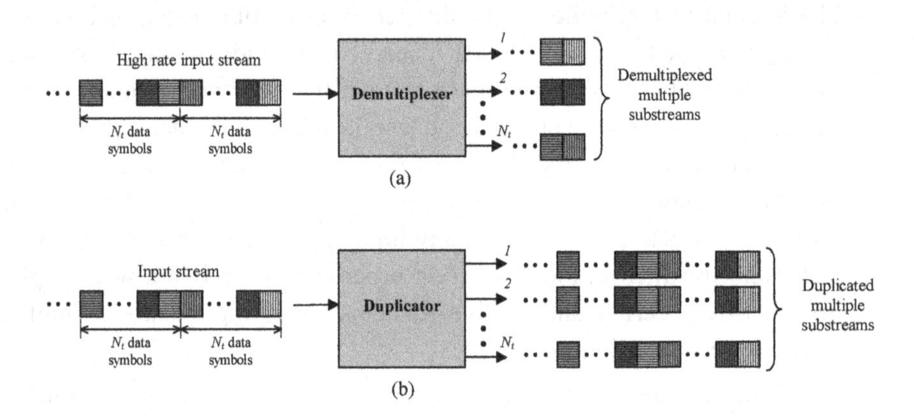

Figure 5.1: (a) Data rate maximisation scheme and (b) diversity maximisation scheme.

The former scheme is performed by demultiplexing a high bit-rate input signal source into multiple data streams and each data stream is then transmitted simultaneously using different transmit antenna. The latter scheme, on the other hand, creates multiple copies of the same input signal source and transmits using multiple antennas to maximise the diversity advantages in the fading channels. In addition, a unique spreading code is assigned to each user to be applied across its transmitting elements. (It is worthwhile to note that the 'same-code assignment' or 'different-code assignment' strategies commonly used in typical CDMA BLAST system [104] can also be adopted in the work that follows.) Hence the modulating information signal due to the jth

transmitting antenna element of the ith user may be written as

$$m_{ij}(t) = \sum_{n=-\infty}^{+\infty} \mathrm{a}_{ij}[n] c_{PN,i}(t - nT_{cs}) \quad (5.1)$$

where $\{\mathrm{a}_{ij}[n], \forall n \in \mathcal{Z}\}$ is the ith user's data symbol transmitted from its jth antenna element, and $c_{PN,i}(t)$ is the pseudo-noise (PN) spreading waveform corresponding to the ith user as described in Equation (2.2).

For a receiver employing an *array-antenna* of N_r sensors, its corresponding spatial array manifold vector associated with the kth path due to the jth transmitting element of the ith user can be defined as $\underline{S}_{ijk} \triangleq \underline{S}(\theta_{ijk})$. If the transmitted signal from the ith user's j^{th} antenna element arrives at the receiver via K_{ij} multipaths, the net received baseband signal vector, in the presence of additive isotropic white Gaussian noise, can be written explicitly as

$$\underline{x}(t) = \sum_{i=1}^{M} \sum_{j=1}^{N_t} \sum_{k=1}^{K_{ij}} \underline{S}_{ijk} \beta_{ijk} \exp(j2\pi f_{ijk} t) m_{ij}(t - \tau_{ijk}) + \underline{\mathrm{n}}(t) \quad (5.2)$$

where $\underline{\mathrm{n}}(t)$ is the complex white Gaussian noise vector; β_{ijk} and τ_{ijk} are respectively the complex fading coefficient and path delay for the kth path associated with the jth transmitting element of the ith user, whilst $\exp(j2\pi f_{ijk} t)$ signifies its corresponding Doppler component with $f_{ijk} = v_{ijk} F_c / c$ (where v_{ijk} is the velocity of the motion towards the receiver, F_c is the carrier frequency, and c is the speed of light) being the Doppler shift frequency. By letting $\mathcal{F}_{ijk}(t) \triangleq \exp(j2\pi f_{ijk} t)$, the net baseband vector representation of the received signal can be rewritten in a more compact form as

$$\underline{x}(t) = \sum_{i=1}^{M} \mathbb{S}_i \operatorname{diag}\left(\underline{\beta}_i \odot \underline{\mathcal{F}}_i(t)\right) \underline{m}_i(t) + \underline{\mathrm{n}}(t) \quad (5.3)$$

where

$$\begin{aligned}
\mathbb{S}_i &= \left[\mathbb{S}_{i1}, \quad \mathbb{S}_{i2}, \quad \ldots, \quad \mathbb{S}_{iN_t}\right] \\
\underline{\beta}_i &= \left[\underline{\beta}_{i1}^T, \quad \underline{\beta}_{i2}^T, \quad \ldots, \quad \underline{\beta}_{iN_t}^T\right]^T \\
\underline{\mathcal{F}}_i(t) &= \left[\underline{\mathcal{F}}_{i1}^T(t), \quad \underline{\mathcal{F}}_{i2}^T(t), \quad \ldots, \quad \underline{\mathcal{F}}_{iN_t}^T(t)\right]^T \\
\underline{m}_i(t) &= \left[\underline{m}_{i1}^T(t), \quad \underline{m}_{i2}^T(t), \quad \ldots, \quad \underline{m}_{iN_t}^T(t)\right]^T
\end{aligned}$$

with $\mathbb{S}_{ij}$, $\underline{\beta}_{ij}$, $\underline{\mathcal{F}}_{ij}(t)$ and $\underline{m}_{ij}(t)$ (for $j = 1, 2, \ldots, N_t$) denoting

$$\begin{aligned}
&\mathbb{S}_{ij} = \left[\underline{S}_{ij1}, \quad \underline{S}_{ij2}, \ldots, \quad \underline{S}_{ijK_{ij}}\right] \\
&\underline{\beta}_{ij} = \left[\beta_{ij1}, \quad \beta_{ij2}, \ldots, \quad \beta_{ijK_{ij}}\right]^T \\
&\underline{\mathcal{F}}_{ij}(t) = \left[\mathcal{F}_{ij1}(t), \quad \mathcal{F}_{ij2}(t), \ldots, \quad \mathcal{F}_{ijK_{ij}}(t)\right]^T \\
&\underline{m}_{ij}(t) = \left[m_{ij}(t - \tau_{ij1}), m_{ij}(t - \tau_{ij2}), \ldots, m_{ij}(t - \tau_{ijK_{ij}})\right]^T
\end{aligned}$$

5.2.2 Discrete Time Doppler-STAR Signal Model

To model the discrete form of the received signal vector, let's first introduce the Spatial-Temporal ARray (STAR) manifold vector due to the kth path from the jth transmitting element of the ith user by modifying Equation (2.16) as

$$\underline{\mathfrak{h}}_{ijk} = \underline{S}_{ijk} \otimes \mathbb{J}^{l_{ijk}} \underline{\mathfrak{c}}_i \tag{5.4}$$

where $l_{ijk} = \lceil \tau_{ijk}/T_s \rceil$ is the corresponding discretised multipath delay. Its Doppler effect can be easily included by extending Equation (5.4) to obtain the following Doppler-STAR manifold vector, given by

$$\underline{\mathfrak{h}}_{ijk}[n] = \underline{S}_{ijk} \otimes \left(\mathbb{J}^{l_{ijk}} \underline{\mathfrak{c}}_i \odot \underline{\mathcal{F}}_{ijk}[n]\right) \tag{5.5}$$

where the time-varying Doppler component $\underline{\mathcal{F}}_{ijk}[n]$ is modelled as

$$\underline{\mathcal{F}}_{ijk}[n] = \begin{bmatrix} 1 \\ \exp(j2\pi f_{ijk} T_s) \\ \exp(j2.2\pi f_{ijk} T_s) \\ \vdots \\ \exp(j2.(L-1)\pi f_{ijk} T_s) \end{bmatrix} \exp(j2n\pi f_{ijk} T_{cs}) \tag{5.6}$$

By taking into consideration the previous, current and next symbols from the jth antenna element of the ith user as $\underline{a}_{ij}[n] = \left[\, a_{ij}[n-1],\, a_{ij}[n],\, a_{ij}[n+1]\,\right]^T$, the discretised representation of the received signal vector $\underline{x}[n]$, containing the Inter-Symbol Interference (ISI) and Multiple-Access Interference (MAI) constituents, can hence be expressed as follows

$$\underline{x}[n] = \sum_{i=1}^{M}\sum_{j=1}^{N_t}\left[\mathbb{H}_{ij}^{\text{prev}}[n]\underline{\beta}_{ij}, \mathbb{H}_{ij}[n]\underline{\beta}_{ij}, \mathbb{H}_{ij}^{\text{next}}[n]\underline{\beta}_{ij}\right]\underline{a}_{ij}[n] + \underline{n}[n] \quad (5.7)$$

where $\underline{n}[n]$ is the sampled noise vector and

$$\begin{aligned}
\mathbb{H}_{ij}[n] &= \left[\underline{\mathfrak{h}}_{ij1}[n], \underline{\mathfrak{h}}_{ij2}[n], \ldots, \underline{\mathfrak{h}}_{ijK_{ij}}[n]\right] && \in \mathcal{C}^{N_r L \times K_{ij}} \\
\mathbb{H}_{ij}^{\text{prev}}[n] &= \left(\mathbb{I}_{N_r} \otimes \left(\mathbb{J}^T\right)^{L/2}\right)\mathbb{H}_{ij}[n] && \in \mathcal{C}^{N_r L \times K_{ij}} \\
\mathbb{H}_{ij}^{\text{next}}[n] &= \left(\mathbb{I}_{N_r} \otimes \mathbb{J}^{L/2}\right)\mathbb{H}_{ij}[n] && \in \mathcal{C}^{N_r L \times K_{ij}}
\end{aligned}$$

Now let's consider the $\mathcal{E}^{\text{th}}$ antenna element of the $\mathcal{D}^{\text{th}}$ user as the desired transmission of interest. By rearranging the terms in Equation (5.7), the net discretised signal vector $\underline{x}[n]$ can thus be rewritten as

$$\begin{aligned}
\underline{x}[n] &= \bar{\mathbb{H}}_{\mathcal{D}\mathcal{E}}[n]\,\mathbb{G}_{\mathcal{D}\mathcal{E}}\,\underline{a}_{\mathcal{D}\mathcal{E}}[n] + \sum_{\substack{j=1 \\ j\neq\mathcal{E}}}^{N_t}\bar{\mathbb{H}}_{\mathcal{D}j}[n]\,\mathbb{G}_{\mathcal{D}j}\,\underline{a}_{\mathcal{D}j}[n] \\
&\quad + \sum_{\substack{i=1 \\ i\neq\mathcal{D}}}^{M}\sum_{j=1}^{N_t}\bar{\mathbb{H}}_{ij}[n]\,\mathbb{G}_{ij}\,\underline{a}_{ij}[n] + \underline{n}[n] \quad (5.8)
\end{aligned}$$

where the first term on the right-hand side of Equation (5.8) denotes the discretised signal vector associated with the $\mathcal{E}^{\text{th}}$ antenna transmission from the desired $\mathcal{D}^{\text{th}}$ user, the second term contains the mutual co-code interferences due to its remaining $(N_t - 1)$ antenna elements' transmissions of the desired user and the third term represents the MAI interferences from the rest of the users together with their associated mutual interferences; the complex fading matrix $\mathbb{G}_{ij} = \mathbb{I}_3 \otimes \underline{\beta}_{ij}$, and the channel matrix $\bar{\mathbb{H}}_{ij}[n] = \left[\,\mathbb{H}_{ij}^{\text{prev}}[n], \mathbb{H}_{ij}[n], \mathbb{H}_{ij}^{\text{next}}[n]\,\right]$.

5.3 Vector Channel Estimation and Reception

5.3.1 Joint Space-Time Channel Estimation

In order to perform an estimation of the channel parameters pertaining to the desired user, a computationally efficient algorithm is devised which requires only the knowledge of the desired user's spreading code sequence. The proposed algorithm, unlike the ones in the previous chapters, does not need a restoration of the desired signal subspace dimensionality. By exploiting the Doppler structural property on the signal data vector, it is noted that the covariance matrix $\mathbb{R}_{xx}$ of the discretised signal vector in Equation (5.8) will preserve and provide a basis for the desired signal subspace, instead of resulting in a dimensional rank deficiency. Similarly, due to the inclusion of the temporal dimension, the number of paths that can be resolved by the algorithm is also no longer constrained by the number of antenna elements available at the array. The algorithm can also be applied for any paths originated from the same source arriving from the same direction (co-directional) or arriving at the same time (co-delay); and even in scenario whereby the paths originated from the different sources of the desired user arriving in the same space and time domains (co-located in space and time). However, for paths of the same source having the same Doppler shift frequencies, singularity in $\mathbb{R}_{xx}$ will occur. For the co-directional cases, robustness has to be put in place within the reception process. Whereas for the co-delay cases, spatial smoothing, similar to that illustrated in Figure 4.3, can be overlaid on top of $\mathbb{R}_{xx}$ to form the smoothed covariance matrix $\mathbb{R}_{\text{Smooth}}$. Having obtained the covariance matrix, the multipaths' space-time channel parameters can then be found jointly by minimising the following MUSIC-type cost function, given as

$$\xi_{\text{ST}}(\theta, l) = \frac{\underline{\mathfrak{b}}_{\mathcal{D}}(\theta, l)^H \, \mathbb{E}_{\text{n,ST}} \mathbb{E}_{\text{n,ST}}^H \, \underline{\mathfrak{b}}_{\mathcal{D}}(\theta, l)}{\underline{\mathfrak{b}}_{\mathcal{D}}(\theta, l)^H \underline{\mathfrak{b}}_{\mathcal{D}}(\theta, l)} \tag{5.9}$$

where $\underline{\mathfrak{b}}_{\mathcal{D}}(\theta, l) = \underline{S}(\theta) \otimes \mathbb{J}^l \underline{\mathfrak{c}}_{\mathcal{D}}$ is the STAR manifold vector with $\underline{\mathfrak{c}}_{\mathcal{D}}$

being related to the spreading code sequence of the desired $\mathcal{D}^{\text{th}}$ user as shown in Equation (2.13), and $\mathbb{E}_{\text{n,ST}}$ is the matrix with columns the noise eigenvectors of $\mathbb{R}_{\text{Smooth}}$.

5.3.2 MUSIC/Analytical Approach to Doppler Estimation

Now by denoting $\widehat{\mathbb{B}}_{ij} \triangleq \left[\widehat{\underline{\mathfrak{b}}}_{ij1}, \widehat{\underline{\mathfrak{b}}}_{ij2}, \ldots, \widehat{\underline{\mathfrak{b}}}_{ijK_{ij}}\right]$, the STAR manifold vector due to the desired user can be reconstructed based on the estimated space-time channel parameters as

$$\widehat{\mathbb{B}}_{\mathcal{D}} = \left[\widehat{\mathbb{B}}_{\mathcal{D}1}, \; \widehat{\mathbb{B}}_{\mathcal{D}2}, \; \cdots, \; \widehat{\mathbb{B}}_{\mathcal{D}N_t}\right] \tag{5.10}$$

In order to suppress the contributions from the MAI and ISI interferences, the received discretised signal vector $\underline{x}[n]$ in Equation (5.8) is thence passed through a novel space-time multipath filter bank to yield

$$\underline{y}[n] = \mathbb{L}^H . \underline{x}[n] \tag{5.11}$$

where $\mathbb{L}$ is the multipath filter bank based on the orthogonal projection of the interference subspace, i.e.

$$\mathbb{L} = \mathbb{P}^{\perp}_{\widehat{\mathbb{B}}_{intf}} \widehat{\mathbb{B}}_{\mathcal{D}} \left(\widehat{\mathbb{B}}^H_{\mathcal{D}} \mathbb{P}^{\perp}_{\widehat{\mathbb{B}}_{intf}} \widehat{\mathbb{B}}_{\mathcal{D}}\right)^{-1} \tag{5.12}$$

where $\mathbb{P}^{\perp}_{\mathbb{K}} = \mathbb{I} - \mathbb{K}(\mathbb{K}^H \mathbb{K})^{-1}\mathbb{K}^H$ is the complementary projection operator of the matrix $\mathbb{K}$; and $\widehat{\mathbb{B}}_{intf} = \left[(\mathbb{I}_{N_r} \otimes (\mathbb{J}^T)^{L/2})\widehat{\mathbb{B}}_{com}, \widehat{\mathbb{B}}^{(\mathcal{D})}_{com},\right.$ $\left.(\mathbb{I}_{N_r} \otimes \mathbb{J}^{L/2})\widehat{\mathbb{B}}_{com}\right]$ in which $\widehat{\mathbb{B}}_{com} = \left[\widehat{\mathbb{B}}_1, \widehat{\mathbb{B}}_2, \ldots, \widehat{\mathbb{B}}_M\right]$ is the estimated composite channel parameters and $\widehat{\mathbb{B}}^{(\mathcal{D})}_{com}$ is the composite matrix $\widehat{\mathbb{B}}_{com}$ with the exclusion of the matrix $\widehat{\mathbb{B}}_{\mathcal{D}}$.

The output of the filter bank, comprising of $\mathcal{O} = \sum_{j=1}^{N_t} K_{\mathcal{D}j}$ branches, is however contaminated with its corresponding Doppler component. This Doppler effect is dominated by the last term on the right-hand side of Equation (5.6) since the temporal difference within a symbol period is comparatively much smaller. Thus for a block of d data symbols under consideration, its Doppler component due to the kth path from the jth transmitting element of the $\mathcal{D}^{\text{th}}$ user can be defined and modelled in terms of its more significant term as $\underline{\Psi}(f_{\mathcal{D}jk}) \triangleq [1, \Psi^1(f_{\mathcal{D}jk}), \Psi^2(f_{\mathcal{D}jk}), \ldots, \Psi^{(d-1)}(f_{\mathcal{D}jk})]^T$ with $\Psi(f_{\mathcal{D}jk}) = \exp(j2\pi f_{\mathcal{D}jk} T_{cs})$. To extract these Doppler parameters, the filter bank output is first pre-processed to remove the phase rotations introduced by the space-time filtered n^{th} symbol:

A. For real data symbols (e.g. BPSK modulation),

$$\underline{y}_{\text{pre}}[n] = \underline{y}[n] \odot \underline{y}[n] \qquad \text{for } n = 1, 2, \ldots, d \qquad (5.13)$$

where d is the full burst length of data symbols being processed by the *array-antenna* system.

B. For complex data symbols (e.g. QPSK modulation),

$$\underline{y}_{\text{pre}}[n] = \underline{y}[n] \odot \underline{\text{a}}[n]^* \qquad \text{for } n = 1, 2, \ldots, d \qquad (5.14)$$

where d is the length of reference data sequence $\underline{\text{a}}[n]$.

With that, the Doppler shift frequencies can then be obtained by performing one-dimensional search for each of the branches (i.e. br1, br2, ..., br$\mathcal{O}$) over the frequency range $f_{\min} \leq f \leq f_{\max}$ using

$$\xi_{\text{D}}(f) = \frac{\underline{\Psi}(\varsigma f)^H \, \mathbb{E}_{\text{n,D}} \, \mathbb{E}_{\text{n,D}}^H \, \underline{\Psi}(\varsigma f)}{\underline{\Psi}(\varsigma f)^H \underline{\Psi}(\varsigma f)} \qquad (5.15)$$

where ς is the factor due to the preprocessing operation in Equations (5.13-5.14) with $\varsigma = 2$ for the case of real data symbols and $\varsigma = 1$ for the case of complex data symbols, and $\mathbb{E}_{\text{n,D}}$ has columns the noise eigenvectors associated with the second order statistics of each pre-processed filter output branch in Equation (5.13) or Equation (5.14).

Alternatively, the estimator can be simplified by evaluating analytically the Doppler shift frequencies across the output branches, $\underline{\widehat{f}} = \left[\widehat{f}_{\text{br1}}, \widehat{f}_{\text{br2}}, \ldots, \widehat{f}_{\text{br}\mathcal{O}}\right]^T$, utilising

$$\underline{\widehat{f}} = \frac{1}{2\pi\varsigma(d-1)T_{cs}} \sum_{n=1}^{d-1} \arg\left(\underline{y}_{\text{pre}}[n+1] \oslash \underline{y}_{\text{pre}}[n]\right) \tag{5.16}$$

5.3.3 Robust MIMO Array Reception

Having found the Doppler shift frequencies, its Doppler effect at the filter bank output in Equation (5.11) can be compensated as follows

$$\underline{z}[n] = \underline{\widehat{\Psi}}[n]^* \odot \underline{y}[n] \tag{5.17}$$

where $\underline{\widehat{\Psi}}[n] = \left[\Psi^n\left(\widehat{f}_{\text{br1}}\right), \Psi^n\left(\widehat{f}_{\text{br2}}\right), \ldots, \Psi^n\left(\widehat{f}_{\text{br}\mathcal{O}}\right)\right]^T$ is the Doppler compensator vector.

For MIMO systems employing the diversity maximisation transmission scheme, the compensated filter bank outputs in Equation (5.17) can then be simply combined to realise the decision statistic for the nth symbol of the desired user

$$b[n] = \underline{w}^H . \underline{z}[n] \tag{5.18}$$

where $\underline{w}$ is the combining weight vector obtained using the principal eigenvector of the autocorrelation matrix of Equation (5.17). On the other hand, for MIMO system with the employment of the data rate maximisation transmission scheme, the compensated filter bank output needs to be partitioned to differentiate those branches belonging to each of the N_t transmitting elements of the desired user. This can be done by passing the compensated output in Equation (5.17), following the decision device, to the *Branch Identification Process*, whereby a cross-correlation is performed with each of the $\mathcal{O}$ branches. The correlated output is then compared with a prespecified threshold value and aggroup together when it exceeds the threshold. Having segregated the branches, the outputs belonging to each of the N_t transmitting elements can then be subsequently combined using Equation (5.18). Note that it is not necessary to assign each and every of the $\mathcal{O}$ branches to all the N_t transmitting elements of the desired user. If the channel parameter of any particular branch is erroneous (incorrect channel estimation) or unidentified (incomplete channel estimation), the *Branch Identification Process* will leave that branch unassigned, thus inducing robustness to the receiver. It is also noticed that in the event whereby two or more paths, arising from the different transmitting elements of the desired user, are co-located in both the space and time domains, the resulted superimposed signal will either be predominated by the strongest signal path and reflected in one of the output branches, or be corrupted and left unassigned by the *Branch Identification Process*. Nevertheless, such occurrences can be minimised by introducing temporal subsymbol separation or increasing the spatial distance separation between the transmitting elements.

Below is a summary of the key step-sequence of the proposed Doppler-STAR MIMO array system operating in a time-varying asynchronous multipath environment:

1. Sample the array output and concatenate the tapped-delay lines (TDL) contents to form the discretised signal vector $\underline{x}[n]$.

2. Form the covariance matrix $\mathbb{R}_{xx}$ of the discretised signal vector and apply the cost function in Equation (5.9) to obtain the joint space-time channel estimation.

3. Based on the estimated channel parameters, compute the multipath filter bank in Equation (5.12) and apply it onto the discretised signal vector as seen in Equation (5.11).

4. Pre-processed the filter bank output as depicted in Equations (5.13-5.14), and perform the cost function in Equation (5.15) or Equation (5.16) to find the associated Doppler shift frequencies.

5. Using the estimated Doppler frequencies, compensate the Doppler effect at the filter bank output as described in Equation (5.17).

   - If the transmission scheme is based on diversity maximisation, simply combine the outputs of the compensated filter bank, as shown in Equation (5.18), using the principal eigenvector of the autocorrelation matrix of Equation (5.17).

   - If the transmission scheme is based on data rate maximisation, apply the *Branch Identification Process* to segregate the output branches associated with each of the transmitting elements of the desired user and combine them using Equation (5.18).

5.4 Simulation Studies

Several representative examples are presented in this section to highlight the key benefits of introducing *array-antenna* technology in typical MIMO systems. Consider a uniform $N_r = 5$ element linear array of half-wavelength spacing operating in the presence of $M = 3$ co-channel BPSK DS-CDMA users, each having $N_t = 2$ transmitting antenna elements, employing the data rate maximisation transmission scheme. Each user is assigned a unique Gold sequence of length $\mathcal{N}_c = 31$ with rectangular chip pulse-shaping. The chip rate is set at $1/T_c = 1.2288$Mchips/s with a carrier frequency of $F_c = 2$GHz. The array is assumed to collect 200 data symbols, with a chip-rate sampler, for processing.

Table 5.1: Users' parameters

User $i = 1$				User $i = 2$				User $i = 3$			
Spreading code vector $\underline{\alpha}_1$				Spreading code vector $\underline{\alpha}_2$				Spreading code vector $\underline{\alpha}_3$			
Path	θ_{ijk}	l_{ijk}	f_{ijk}	Path	θ_{ijk}	l_{ijk}	f_{ijk}	Path	θ_{ijk}	l_{ijk}	f_{ijk}
Antenna element $j = 1$				Antenna element $j = 1$				Antenna element $j = 1$			
$k = 1$	40°	$8T_c$	30Hz	$k = 1$	30°	$10T_c$	-90Hz	$k = 1$	20°	$15T_c$	40Hz
$k = 2$	50°	$18T_c$	100Hz	$k = 2$	70°	$25T_c$	60Hz	$k = 2$	60°	$4T_c$	-80Hz
$k = 3$	70°	$25T_c$	-160Hz	$k = 3$	80°	$20T_c$	170Hz	$k = 3$	60°	$7T_c$	-10Hz
$k = 4$	90°	$18T_c$	100Hz	$k = 4$	80°	$21T_c$	30Hz	$k = 4$	60°	$10T_c$	5Hz
$k = 5$	100°	$12T_c$	0Hz	$k = 5$	100°	$10T_c$	-110Hz	Antenna element $j = 2$			
Antenna element $j = 2$				$k = 6$	110°	$11T_c$	-100Hz	$k = 1$	100°	$3T_c$	180Hz
$k = 1$	60°	$15T_c$	150Hz	Antenna element $j = 2$				$k = 2$	110°	$20T_c$	-2Hz
$k = 2$	90°	$5T_c$	-80Hz	$k = 1$	80°	$5T_c$	-60Hz	$k = 3$	130°	$11T_c$	110Hz
$k = 3$	90°	$20T_c$	100Hz	$k = 2$	90°	$5T_c$	-60Hz	$k = 4$	130°	$12T_c$	-50Hz
$k = 4$	120°	$10T_c$	0Hz	$k = 3$	120°	$28T_c$	150Hz	$k = 5$	140°	$8T_c$	90Hz
$k = 5$	130°	$15T_c$	-120Hz	$k = 4$	150°	$25T_c$	0Hz	$k = 6$	160°	$18T_c$	-170Hz

5.4.1 Performance of Doppler-Space-Time Algorithm

Let's take user 1 as the desired user having an input SNR of 20dB; while the rest of the interferers each constituting an interference ratio of 20dB (i.e. near-far problem). All 3 users are assumed to have 10 multipaths each, with their parameters as listed in Table 5.1. The

Doppler spread is set at 200Hz which corresponds to a maximum speed of 108km/h. By partitioning the array into 2 overlapping 4-element subarrays for spatial smoothing, it is seen from Figure 5.2 that all the 10 multipaths, associated with the two transmitting elements of the desired user, can be identified/estimated successfully using the proposed algorithm. Notice that the algorithm can still operate even when the desired user's paths are co-located in both (i) the code and space domains, (ii) the code and time domains, or (iii) the space and time domains. In addition to that, the number of multipaths that can be resolved by the algorithm is also not constrained by the number of antennas available in the array.

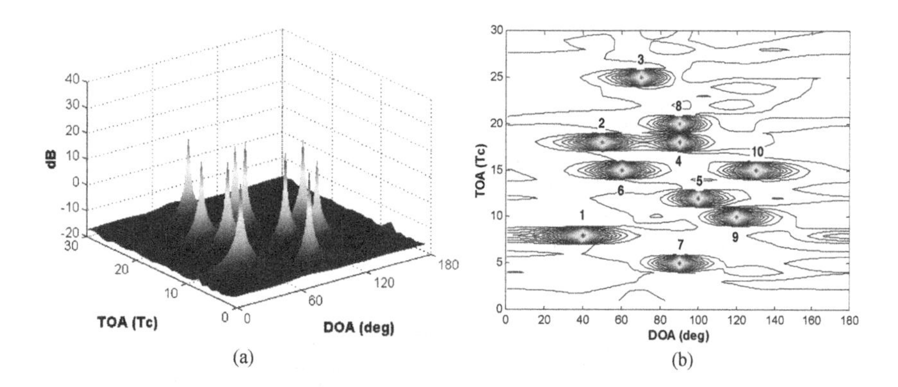

Figure 5.2: Joint space-time spectrum of all multipaths due to the desired user.

The space-time filtered output is then preprocessed as described in Equation (5.13) with $d = 200$ data symbols. By setting the search step size as 1/10Hz, the Doppler-MUSIC spectrum due to the desired user is plotted as depicted in Figure 5.3. Its results are compared with its corresponding Doppler-Analytical estimator and tabulated as shown in Table 5.2. It is therefore apparent that the Doppler shift frequencies associated with all the 10 multipaths, be it identical or zero, can be correctly estimated using either the Doppler-MUSIC or Doppler-Analytical approaches.

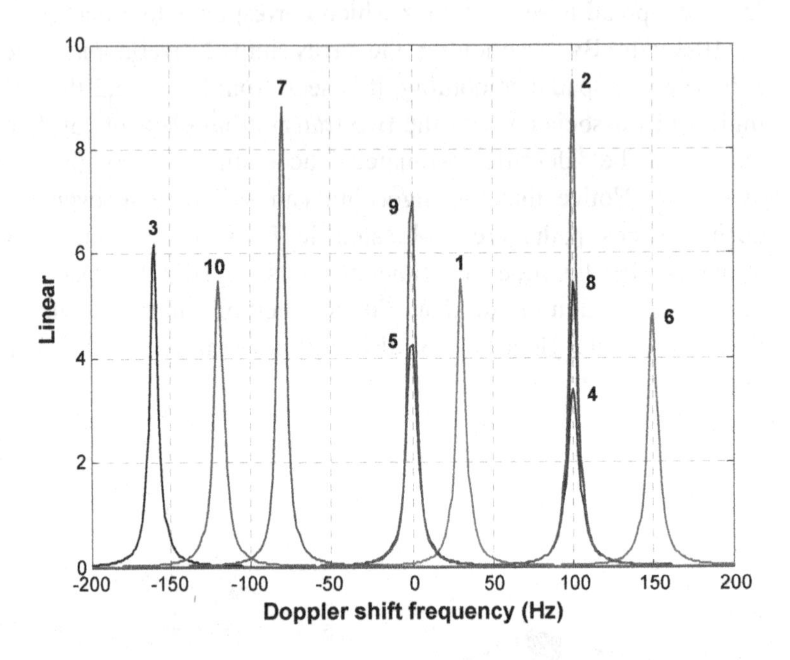

Figure 5.3: Doppler shift frequency spectrum of the desired user.

Table 5.2: Doppler shift frequency estimation of desired user $i = 1$

	Parameter to be estimated	Doppler shift frequency f_{1jk} (Hz)				
Antenna element $j = 1$	Associated k^{th} path component	1	2	3	4	5
	Doppler-MUSIC estimation	30.1	99.8	-159.8	100.1	0.0
	Doppler-Analytical estimation	30.3	100.6	-160.5	99.8	0.3
Antenna element $j = 2$	Associated k^{th} path component	1	2	3	4	5
	Doppler-MUSIC estimation	150.1	-80.1	99.7	-0.2	-120.0
	Doppler-Analytical estimation	150.4	-79.5	100.8	0.6	-120.1

The multipaths belonging to each of the two transmitting elements of the desired user are then singled out by applying the *Branch Identification Process* to the compensated filter bank output. Its cross-correlation with the first output branch is as illustrated in Table 5.3. By setting a prespecified threshold value of, say 0.5, for comparison, the multipaths due to the each of the antenna elements can be clearly identified from the table. Having segregated the paths, Figure 5.4 depicts the performance of the proposed Doppler-STAR receiver as compared with conventional ST decorrelating detector and 2D RAKE receiver, with the latter two assuming full knowledge of the channel. The proposed receiver is thus seen to be more tolerant with the multipath Doppler spread, whilst the decorrelating detector and RAKE receiver deteriorate drastically at the onset of the spread. Note that the gradual degradation of the proposed receiver's output SNIR is as a result of the Doppler variation effect within a symbol period which sets in at about 150Hz. Nevertheless, the proposed receiver still provides substantial performance gain in the Doppler spread range under consideration.

Table 5.3: Branch Identification Process of user $i = 1$

(Ant. j, Path k)	(1,1)	(1,2)	(1,3)	(1,4)	(1,5)	(2,1)	(2,2)	(2,3)	(2,4)	(2,5)
Correlation	1.000	0.999	0.999	1.000	0.999	0.025	0.025	0.024	0.025	0.024

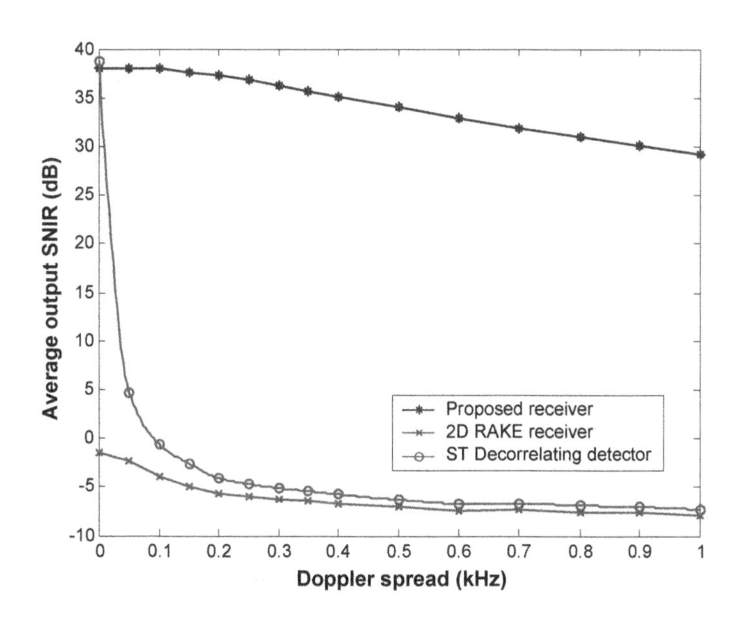

Figure 5.4: Output SNIR performance versus Doppler spread.

5.4.2 Comparison of Doppler Estimation Approaches

Now let's examine and compare the performance between the Doppler-MUSIC approach and the Doppler-Analytical approach of the estimation algorithm. The RMSE of the Doppler estimates using both the approaches are averaged over 200 Monte Carlo runs and plotted with respect to the multipath Doppler spread as illustrated in Figure 5.5. Note that the bottom two curves in the figure correspond to real data modulation with $d = 200$; whereas the top two curves correspond to complex data modulation with $d = 50$. As expected, since the length of the data block d being processed by the Doppler estimators is longer for the real case, the accuracy of its Doppler estimates is therefore much better when compared with the complex one. Nevertheless in both cases, the MUSIC approach is seen to give a relatively lower RMSE than that using the analytical approach. In spite of that, the analytical Doppler estimator still provides a reasonably good approximation. And as similar to the SNIR study in Figure 5.4, the effect of the Doppler variation within a symbol period is seen to set in after a spread of about 150Hz.

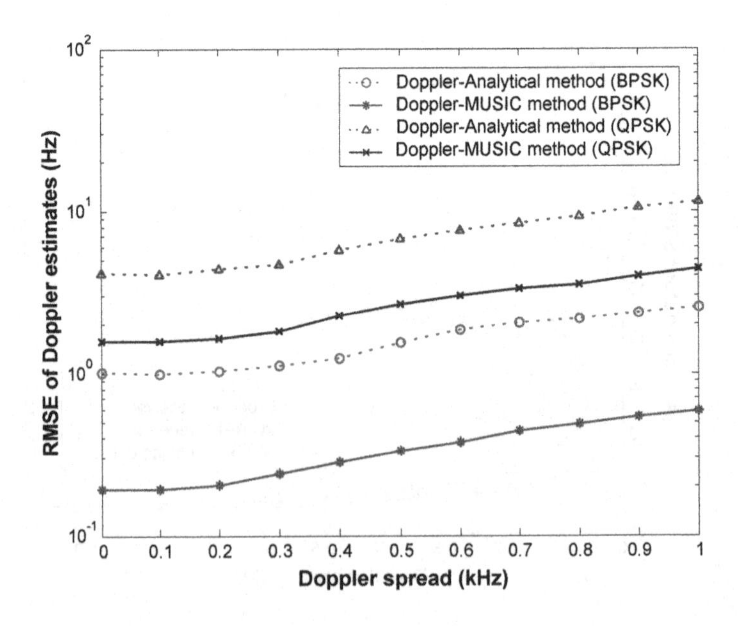

Figure 5.5: RMSE of Doppler-MUSIC and Doppler-Analytical estimations.

Next, let's take a closer look at the impact on the accuracy of the estimates in response to the length of the reference data sequence for the complex data modulation case. As anticipated, its estimates, as shown in Figure 5.6, improve with the length of the reference sequence, with the MUSIC approach faring better than its analytical counterpart. Nonetheless, both the estimates' accuracy decline rather rapidly as the reference sequence falls below the length of about 25 data symbols.

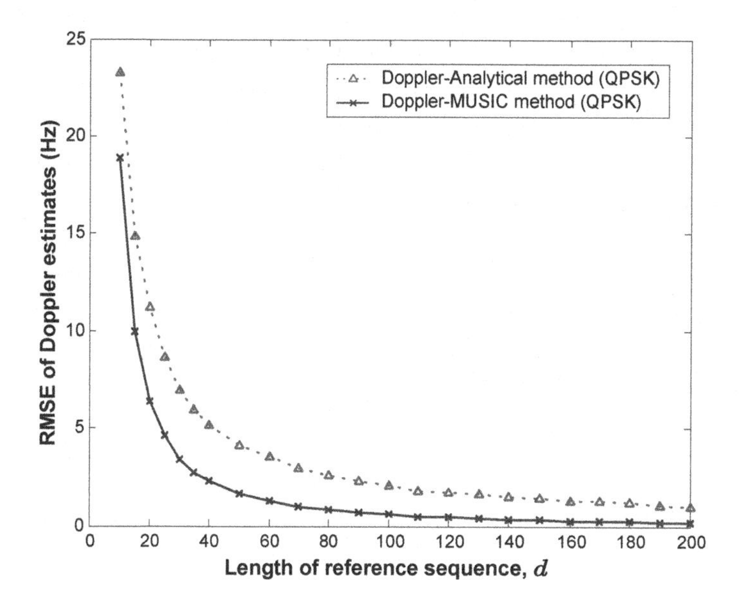

Figure 5.6: Doppler estimation accuracy with respect to reference sequence length.

5.4.3 Investigation of Near-Far Resistant Capability

Using the same scenario as in Table 5.1, the near-far resistant capability of the proposed receiver is investigated in greater detail as depicted in Figure 5.7. The power level of the two interferers is varied with respect to that of the desired user, with each constituting a near-far ratio (NFR) of between -20dB and 100dB. Unlike the proposed receiver which has a relatively consistent average output SNIR till a NFR value of about 40dB, the ST decorrelating detector and 2D RAKE receiver as shown in Figure 5.7a are observed to be deteriorating with increasing amount of NFR. It is also evident from the BER plot in Figure 5.7b that the proposed receiver is more insusceptible to the near-far problem; whereas the error floors of the decorrelating detector and RAKE receiver shoot up shortly after the desired user's signal power drops below the interference power level.

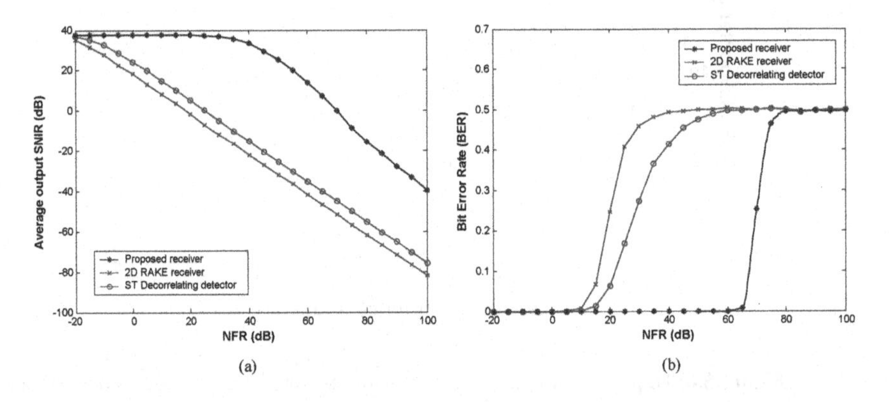

Figure 5.7: Receivers' near-far resistant performance.

5.5 Summary

A blind MIMO array receiver based on a computationally efficient near-far resistant channel estimation technique is presented for time-varying asynchronous multipath DS-CDMA system. Without the need of exhausting the number of available spreading codes, the proposed receiver is applicable in either the data rate maximisation or diversity maximisation scheme. The potential benefit of incorporating the *array-antenna* system is demonstrated by the near-far resistant Doppler-space-time estimation algorithm which does not require the need of any power control or the knowledge of the channel, as are readily assumed in most MIMO systems. Due to the inclusion of the *Branch Identification Process* in the detection procedure, the receiver is robust to erroneous or incomplete channel estimation even in the event whereby the multipaths are co-located in the space, time and code domains. The number of resolvable multipaths is also no longer limited by the number of antenna elements available in the array.

5.6 Recommendations

The above proposed Doppler-space-time estimation and reception algorithm is applicable for a wide range of Doppler frequencies, be it small or large, provided that the spatial, temporal and Doppler channel parameters remain rather quasi-stationary during the observation window period, as have assumed in the simulation studies. However its space-time estimation algorithm is unable to detect co-directional signal paths, originated from the same source, having the same Doppler shift frequencies; which hence necessitates the need of robustness inbuilt in the reception process.

To resolve such undetectable occurrences, an alternative approach to the Doppler-space-time estimation process will be examined in Section 6.1. Unlike the above, its space-time channel estimation algorithm is able to detect all instances of co-directional or co-delay signal paths, as long as the effect from the intra-symbol Doppler contribution stays trivial. And its Doppler shift frequency estimation algorithm, which is formulated based on the intra-symbol Doppler component, is much simpler; albeit its performance, as anticipated, is not as comparable as that of the above which is formulated based on the more dominant inter-symbol Doppler component.

Hence depending on the application, the following combinations of the estimation approach can be considered to improve/enhance the performance of the Doppler-space-time estimation process.

- For large Doppler applications, the space-time estimation approach in Equation (5.9) can be used in conjunction with the Doppler estimation approach in Equation (6.6) to take advantage of reusing the same noise eigenvectors matrix used in both of the expressions. The resulted algorithm is thus less computational but the robustness requirement needs to be built into the reception process.
- For small Doppler applications, the space-time estimation approach in Equation (6.5) can be integrated together with the Doppler estimation approach in Equation (5.15) to avoid any undetectable occurrences co-located in either (a) the same space, Doppler and code domains or (b) the same time, Doppler and code domains. Robustification of the reception process is therefore not necessary to be inbuilt into the system.
- For wide-ranging Doppler applications, the Doppler-space-time channel estimation algorithm, as described herein in Section 5.3, is deemed to be the best approach.

CHAPTER 6

Estimation and Reception Application Examples

Having seen the different estimation and reception algorithms, let's explore the application of the algorithmic framework adapting to the new frontier of research. Three representative examples are included herein to aid in a deeper understanding as well as to gain a new perspective of the algorithm. The first takes a further examination into the space-time Doppler algorithm in a single input multiple output sensor network system; the second introduces the use of the transmit diversity concept integrated with the estimation and reception process; and the third exploits the employment of polarisation multiplexing in a multiple input multiple output system. These are some of the examples which illustrate the application of the algorithmic tools devised in accordance to the context of the environment.

6.1 SIMO Sensor Network Array System

The deployment of wireless sensor network in an uncertain or harsh environment has gained much interest in the recent years, especially in the healthcare, military, security, or natural-catastrophe applications. Of particular importance is the transmission link from the sensor nodes to the central receiver which is acting as the information sink for the data gathered from the network. The link could suffer from several detrimental impairments as a result of (a) multipath propagation effect induced by multiple signal reflection, refraction, diffraction and scattering, (b) relative motion due to the employment of non-stationary central receiver and/or moving sensor nodes, and (c) near-far problem caused by near-distant sensor node or weak signal reception in the event of sensor failure or weakened battery power [115]. This work aims to address the above issues in a single-hop wireless sensor network operating in an asynchronous time-varying frequency-selective DS-CDMA channel. A novel Doppler-space-time estimation and reception algorithm is thus devised, for Single Input Multiple Output (SIMO) sensor network, which does not impose additional processing burden onto the sensor nodes. Multiple antenna elements are adopted at the central receiver to form a Spatial-Temporal ARray (STAR) system which allows superresolution beamforming to be directed in a specific space and time domain. By exploiting the spatial-temporal properties of the channel, an extra layer of interference cancellation and new ways of handling unwanted channel effects can hence be developed. An extra form of diversity, in addition to the diversities in the space and time domains, can also be derived from the often-regarded detrimental Doppler frequency domain [113]. The proposed algorithm is near-far resistant and does not require the knowledge of the channel. Unlike in previous chapter, it is also capable of operating in an environment with or without the presence of the Doppler effect.

6.1.1 SIMO Sensor Network Signal Model

Consider a single-hop SIMO wireless sensor network, comprising of a set of M sensor nodes, operating in an asynchronous DS-CDMA multipath channel. A random spreading code, with a length of $\mathcal{N}_c$, is assigned to each of the sensor nodes. Suppose the transmitted signal from the ith sensor node arrives at the central receiver via K_i multipaths. For a receiver employing an array of N antenna elements, the discretised received signal vector, after passing through a bank of tapped-delay lines of length L, with a sampling period T_s, can therefore be written as

$$\underline{x}[n] \quad = \quad \sum_{i=1}^{M} \bar{\mathbb{H}}_i[n]\mathbb{G}_i\underline{\mathrm{a}}_i[n] + \underline{\mathrm{n}}[n] \tag{6.1}$$

where $\underline{\mathrm{n}}[n]$ is the sampled additive white Gaussian noise vector, $\underline{\mathrm{a}}_i[n] = \left[\, \mathrm{a}_i[n-1],\, \mathrm{a}_i[n],\, \mathrm{a}_i[n+1] \,\right]^T$ is the previous, current and next symbol contributions, $\mathbb{G}_i = \mathbb{I}_3 \otimes \underline{\beta}_i$ consists of the multipaths' complex fading coefficients; and $\bar{\mathbb{H}}_i[n] = \left[\, \left(\mathbb{I}_N \otimes (\mathbb{J}^T)^{L/2}\right) \mathbb{H}_i[n],\, \mathbb{H}_i[n],\right.$ $\left.\left(\mathbb{I}_N \otimes \mathbb{J}^{L/2}\right) \mathbb{H}[n] \,\right]$ is the channel matrix in which $\mathbb{J}$ (or $\mathbb{J}^T$) is a time down-shift (or up-shift) matrix and $\mathbb{H}_i[n] = \left[\, \underline{\mathfrak{h}}_{i1}[n],\, \underline{\mathfrak{h}}_{i2}[n],\, \ldots,\, \underline{\mathfrak{h}}_{iK_i}[n]\right]$ with $\underline{\mathfrak{h}}_{ij}[n]$ being the Doppler-STAR manifold vector defined as

$$\underline{\mathfrak{h}}_{ij}[n] \quad = \quad \underline{\mathfrak{h}}_{ij}\exp(j2n\pi f_{ij}T_{cs}) \tag{6.2}$$

The exponential term denotes the inter-symbol Doppler constituent with f_{ij} being its associated Doppler shift frequency, and $\underline{\mathfrak{h}}_{ij}$ is the manifold vector encompassing the spatial, temporal and intra-symbol Doppler effects modelled as:

$$\underline{\mathfrak{h}}_{ij} \quad = \quad \underline{S}_{ij} \otimes \left(\mathbb{J}^{l_{ij}}\underline{\mathfrak{c}}_i \odot \underline{\Upsilon}_{ij}\right) \tag{6.3}$$

where $\underline{S}_{ij} \triangleq \underline{S}(\theta_{ij})$ is the spatial array manifold vector, l_{ij} is the discretised multipath delay, $\underline{\mathfrak{c}}_i$ is related to the ith sensor node's spreading sequence, and $\underline{\Upsilon}_{ij} \triangleq \underline{\Upsilon}(f_{ij})$ is the intra-symbol Doppler constituent given as $\underline{\Upsilon}_{ij} = \left[\, 1,\, \Upsilon^1(f_{ij}),\, \Upsilon^2(f_{ij}),\, \ldots,\, \Upsilon^{(L-1)}(f_{ij}) \,\right]^T$ in which $\Upsilon(f_{ij}) = \exp(j2\pi f_{ij}T_s)$.

6.1.2 Blind Estimation and Reception

The received signed vector is however contaminated by the Inter-Symbol Interference (ISI), Multiple-Access Interference (MAI) and noise components. To perform the space-time channel estimation pertaining to the desired sensor node, the received discretised signal vector $\underline{x}[n]$ can be preprocessed by the desired $\mathcal{D}^{\text{th}}$ sensor node's preprocessor $\mathbb{Z}_{\mathcal{D}} = \mathbb{I}_N \otimes \left(\text{diag}(\underline{\tilde{\mathfrak{c}}}_{\mathcal{D}})^{-1}\mathbb{F}\right)$ as described in Equation (3.19). (It is worth noting that the inverse operation within the preprocessor $\mathbb{Z}_{\mathcal{D}}$ can only be valid if and only if the sum of its random spreading code sequence is non-zero, that is $\sum_{m=0}^{\mathcal{N}_c-1} \alpha_{\mathcal{D}}[m] \neq 0$.)

Since the contribution from the intra-symbol Doppler effect is not as significant as compared with that from the inter-symbol Doppler effect, the preprocessing operation is able to effectively isolate the desired signal component and transform its temporal domain to conform to a Vandermonde structure. Similarly as in previous study, it is not necessary to restore the desired signal subspace dimensionality of the preprocessed signal vector. The Doppler structural property of the signal data vector will preserve and provide a basis for the desired signal subspace. This can also be applied for any co-delay or co-directional paths originated from the same sensor node, or any co-located (same space and time domains) paths originated from the different sensor nodes. But for paths of the same sensor node having the same Doppler shift frequencies arriving in the same space or time domain, singularity in the covariance matrix will occur. This however can be resolved by applying spatial-temporal smoothing on the preprocessed signal vector, with the spatial smoothing portion restoring the co-delay cases and the temporal smoothing portion restoring the co-directional cases. The concept is similar to the well-known spatial smoothing technique described in [80]. But to do this in a more effective manner, a joint space-time smoothing technique is proposed herein. Let's first reshape the preprocessed signal vector $\underline{y}[n] \in \mathcal{C}^{NL\times 1}$ to a matrix $\mathbb{Y}[n] \in \mathcal{C}^{N\times L}$ as shown in Figure 6.1. By extracting a set of $e \times d$ submatrices across the

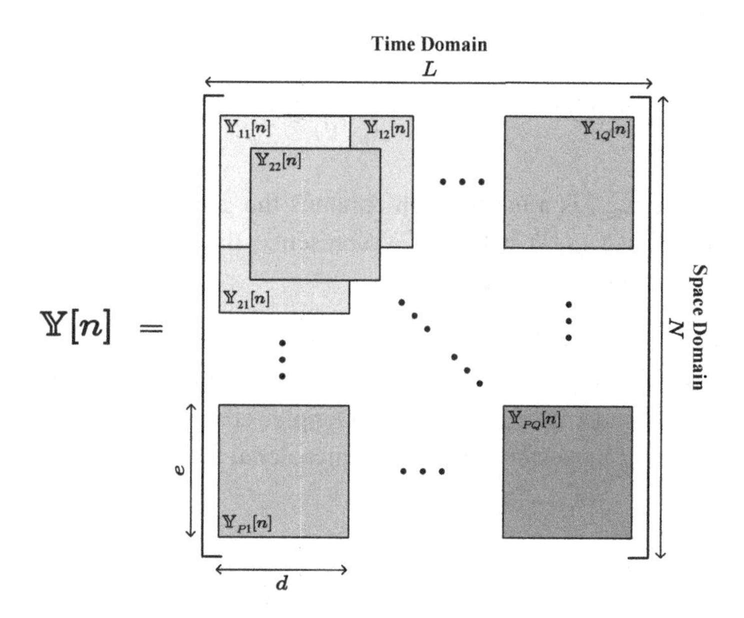

Figure 6.1: Joint spatial-temporal smoothing technique.

space and time dimensions, the spatial-temporal smoothed covariance matrix can hence be found as follows

$$\mathbb{R}_{\text{STsmooth}} = \frac{1}{PQ} \sum_{u=1}^{P} \sum_{v=1}^{Q} \mathbb{R}_{\underline{y}_{uv}} \tag{6.4}$$

where $\mathbb{R}_{\underline{y}_{uv}}$ is the covariance matrix obtained from the subvector $\underline{y}_{uv}[n] = \text{vec}(\mathbb{Y}_{uv}[n])$ with $\text{vec}(\bullet)$ being the row-wise vectorisation operator. Note that for $e < N$ and $d = L$, spatial smoothing will be performed without temporal smoothing; likewise for $d < L$ and $e = N$, only temporal smoothing will be carried out without spatial smoothing.

With that, the space-time channel parameters of the multipaths can thus be identified using a two-dimensional MUSIC-type cost function as shown below:

$$\xi_{\mathrm{ST}}(\theta, l) = \frac{\left(\underline{S}(\theta) \otimes \underline{\Phi}_d^l\right)^H \mathbb{E}_{\mathrm{n,ST}} \mathbb{E}_{\mathrm{n,ST}}^H \left(\underline{S}(\theta) \otimes \underline{\Phi}_d^l\right)}{\left(\underline{S}(\theta) \otimes \underline{\Phi}_d^l\right)^H \left(\underline{S}(\theta) \otimes \underline{\Phi}_d^l\right)} \quad (6.5)$$

where $\mathbb{E}_{\mathrm{n,ST}}$ is a matrix with columns the generalised noise eigenvectors of $(\mathbb{R}_{\mathrm{STsmooth}}, \mathbb{D})$ whilst $\mathbb{D}$ represents the spatial-temporal-smoothed diagonal matrix $\mathbb{Z}_{\mathcal{D}}\mathbb{Z}_{\mathcal{D}}^H$.

Having obtained the Direction-of-Arrival (DOA) $\widehat{\theta}_{\mathcal{D}j}$ and Time-of-Arrival (TOA) $\widehat{l}_{\mathcal{D}j}$ of the jth signal path associated with the desired $\mathcal{D}^{\text{th}}$ sensor node, its corresponding Doppler channel parameter can therefore be found by performing a one-dimensional search as follows

$$\xi_{\mathrm{D}}(f) = \frac{\left(\underline{S}(\widehat{\theta}_{\mathcal{D}j}) \otimes (\mathbb{J}^{\widehat{l}_{\mathcal{D}j}} \underline{\mathfrak{c}}_{\mathcal{D}} \odot \underline{\Upsilon}(f))\right)^H \mathbb{E}_{\mathrm{n,D}} \mathbb{E}_{\mathrm{n,D}}^H \left(\underline{S}(\widehat{\theta}_{\mathcal{D}j}) \otimes (\mathbb{J}^{\widehat{l}_{\mathcal{D}j}} \underline{\mathfrak{c}}_{\mathcal{D}} \odot \underline{\Upsilon}(f))\right)}{\left(\underline{S}(\widehat{\theta}_{\mathcal{D}j}) \otimes (\mathbb{J}^{\widehat{l}_{\mathcal{D}j}} \underline{\mathfrak{c}}_{\mathcal{D}} \odot \underline{\Upsilon}(f))\right)^H \left(\underline{S}(\widehat{\theta}_{\mathcal{D}j}) \otimes (\mathbb{J}^{\widehat{l}_{\mathcal{D}j}} \underline{\mathfrak{c}}_{\mathcal{D}} \odot \underline{\Upsilon}(f))\right)} \quad (6.6)$$

where $\mathbb{E}_{\mathrm{n,D}}$ is a matrix with columns the noise eigenvectors of the second order statistics of $\underline{x}[n]$. Similarly, the Doppler estimator can also be applied for any co-delay or co-directional scenarios, except for instances of signal paths co-located in the same time, code and Doppler domains in which spatial smoothing has to be performed to reconstruct the dimensionality.

From the estimated spatial, temporal and Doppler parameters, the signal paths pertaining to the desired $\mathcal{D}^{\text{th}}$ sensor node can now be segregated by suppressing the ISI and MAI interferences from the received signal vector $\underline{x}[n]$, that is

$$\underline{z}[n] = \operatorname{diag}(\underline{\widehat{\Psi}}[n]^*).(\mathbb{L}^H.\underline{x}[n]) \quad (6.7)$$

with the vector $\underline{\widehat{\Psi}}[n] = \left[\Psi^n(\widehat{f}_{\mathcal{D}1}), \Psi^n(\widehat{f}_{\mathcal{D}2}), \ldots, \Psi^n(\widehat{f}_{\mathcal{D}K_{\mathcal{D}}})\right]^T$ being the inter-symbol Doppler shift frequency compensator in which $\Psi^n(\widehat{f}_{ij}) = \exp\left(j2n\pi\widehat{f}_{ij}T_{cs}\right)$; and the matrix $\mathbb{L}$ is the multipath filter bank based on the projection of the noise subspace spanned by the interferences, i.e.

$$\mathbb{L} \quad = \quad \mathbb{P}_{\mathbb{E}_n} \widehat{\mathbb{H}}_{\mathcal{D}} \left(\widehat{\mathbb{H}}_{\mathcal{D}}^{H} \mathbb{P}_n \widehat{\mathbb{H}}_{\mathcal{D}} \right)^{-1} \tag{6.8}$$

where $\mathbb{P}_{\mathbb{E}_n} = \mathbb{E}_n (\mathbb{E}_n^H \mathbb{E}_n)^{-1} \mathbb{E}_n^H$ is the projection operator matrix onto the subspace spanned by the noise eigenvectors of $\mathbb{R}_{intf} = \mathbb{R}_{xx} - \widehat{\mathbb{H}}_{\mathcal{D}} \widehat{\mathbb{H}}_{\mathcal{D}}^H$ in which $\mathbb{R}_{xx}$ is the covariance matrix of $\underline{x}[n]$ and $\widehat{\mathbb{H}}_{\mathcal{D}} \triangleq [\underline{\widehat{\mathfrak{h}}}_{\mathcal{D}1}, \underline{\widehat{\mathfrak{h}}}_{\mathcal{D}2}, \ldots, \underline{\widehat{\mathfrak{h}}}_{\mathcal{D}K_{i\mathcal{D}}}]$.

To realise the nth symbol's decision statistic of the desired sensor node, the segregated filter bank outputs can hence be combined to yield

$$b[n] \quad = \quad \underline{w}^H . \underline{z}[n] \tag{6.9}$$

where $\underline{w}$ is the combining weight vector obtained using the principal eigenvector of the autocorrelation matrix of Equation (6.7).

6.1.3 Performance Analysis

Consider a single-hop sensor network comprising of a central receiver having a uniform $N = 5$ element linear array of half-wavelength spacing operating in the presence of $M = 3$ asynchronous DS-CDMA sensor nodes with each assigned a randomly-generated code sequence of length $\mathcal{N}_c = 28$. The carrier frequency is set at $F_c = 2.4$GHz, and the associated Doppler spread is preset at 150Hz to account for the relative motion between the central receiver and the sensor nodes. The receiver collects blocks of 100 data symbols with a chip-rate sampling period of $T_s = 8.14\mu$sec.

Suppose the desired sensor node has a SNR of 20dB, but suffers a near-far effect with each of the remaining sensor nodes constituting an interference ratio of about 80 times higher than that of the desired sensor node. Each of the nodes is assumed to have about 6 to 8 multipaths, with the desired sensor node having 7 signal paths originated from its transmitting element. By setting $e = 4$ for spatial smoothing and $d = 55$ for temporal smoothing, all the multipaths associated with the desired sensor node, including the co-delay and co-directional signal paths, can hence be identified in the space-time spectrum plot as shown in Figure 6.2. Likewise, by partitioning the antenna array into 2 overlapping 4-element subarrays for spatial smoothing, its corresponding Doppler shift frequencies, be it identical or zero, can be successfully estimated as illustrated by the Doppler spectrum plot in Figure 6.3. Notice that the proposed estimation algorithm can operate even when the signal paths of the desired sensor node are co-located in either the (i) time and Doppler domains, or (ii) space and Doppler domains. The number of resolvable multipaths is also not constrained by the number of antennas elements available in the array.

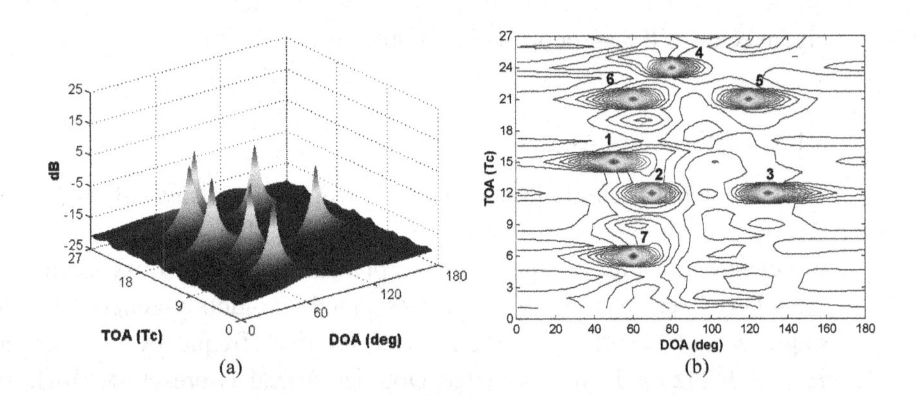

Figure 6.2: Space-time spectrum of the desired sensor node.

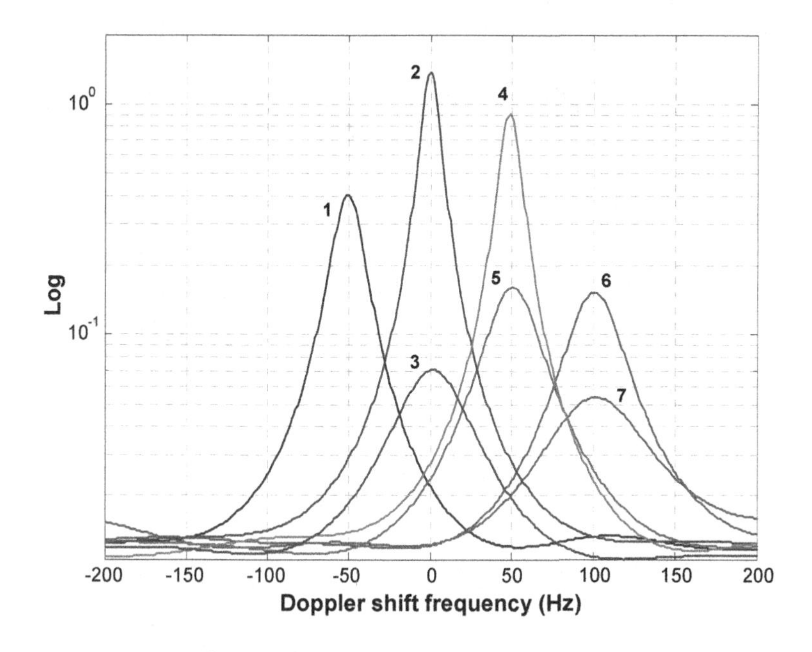

Figure 6.3: Doppler frequency spectrum of desired sensor node.

Figure 6.4 depicts the performance of the proposed blind Doppler-STAR receiver as compared with conventional space-time decorrelating detector and 2D RAKE receiver, with the latter two assuming complete knowledge of the channel. It is hence apparent that the proposed receiver, in the absence of prior channel knowledge, is observed to be better in performance, with its operation more tolerant with the Doppler effect. In fact, the proposed receiver is able to operate irregardless of whether the central receiver and/or sensor nodes are moving or stationary in position.

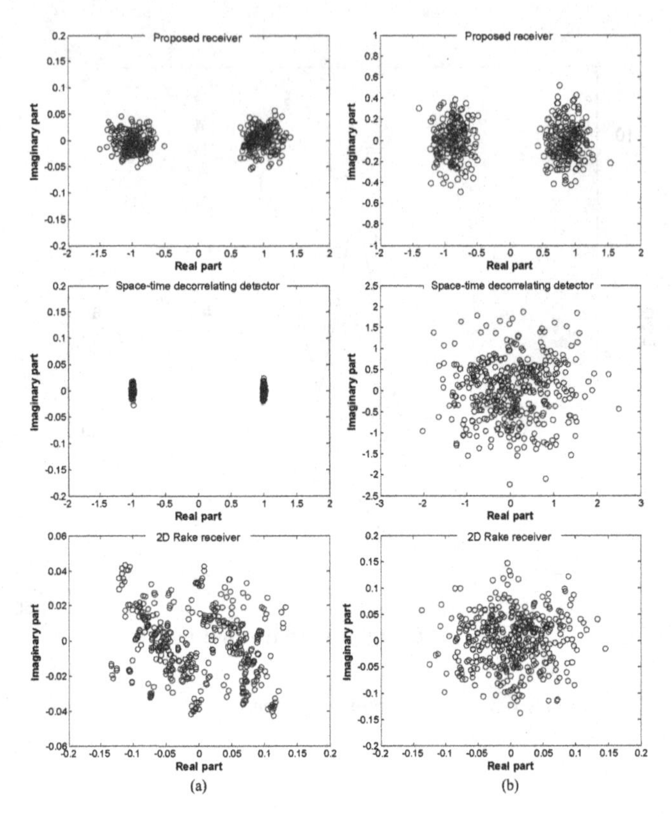

Figure 6.4: Constellation diagram of proposed receiver, space-time decorrelating detector and 2D RAKE receiver with (a) Doppler spread = 0Hz and (b) Doppler spread = 150Hz.

6.1.4 Concluding Summary

In this work, a novel Doppler-space-time estimation and reception algorithm is proposed for single-hop SIMO sensor network operating in a harsh environmental condition. The algorithm can perform irregardless of whether the central receiver and/or sensor nodes are in motion or in static position. It requires no knowledge of the channel and is resistant against MAI, ISI and other near-far interferences as a result of strong near-distant sensor nodes or weakened transmitted signal power. The improvement does not impose additional processing burden on the sensor nodes, but instead on the central receiver itself.

6.2 MIMO OTD Array System

Orthogonal transmit diversity (OTD) scheme, proposed by Motorola [116], was adopted as one of the open loop diversity techniques in cdma2000 standard [117]. The scheme transmits its signals via multiple antennas, and ensures orthogonality between its transmissions by means of Walsh-Hadamard coding [105]. However most works on OTD frequently assume knowledge of the channel in a flat fading environment. In this study, a novel asynchronous DS-CDMA MIMO array receiver, exploiting the transmission methodology of OTD scheme, is proposed in a slow time-varying frequency-selective environment. By incorporating the *array-antenna* technology at the front-end of the receiver, the proposed approach is blind and near-far resistant. Furthermore, it does not exhaust the limited number of spreading codes available, as in the case of conventional multi-code MIMO designs. The proposed formulation is also applicable in both the uplink and the downlink.

6.2.1 MIMO OTD Signal Model

Consider a MIMO array system comprises of an N_r elements *array-antenna* receiver operating in the presence of M asynchronous users with each employing the OTD transmission scheme across its $N_t = 2$ transmitting antenna elements. The corresponding output symbols due to the ith user's differential OTD encoder, as depicted in Figure 6.5, can be described as

$$\begin{bmatrix} \mathrm{a}_{i1} & \mathrm{a}_{i1} \\ \mathrm{a}_{i2} & -\mathrm{a}_{i2} \end{bmatrix} \tag{6.10}$$

where each row denotes the two consecutive symbols transmitted at each antenna, and each column denotes the odd and even symbols across the antennas transmitted at each time slot. The output symbols are then spread by the ith user's code sequence $\underline{\alpha}_i$ of length $\mathcal{N}_c$. Note that it is not

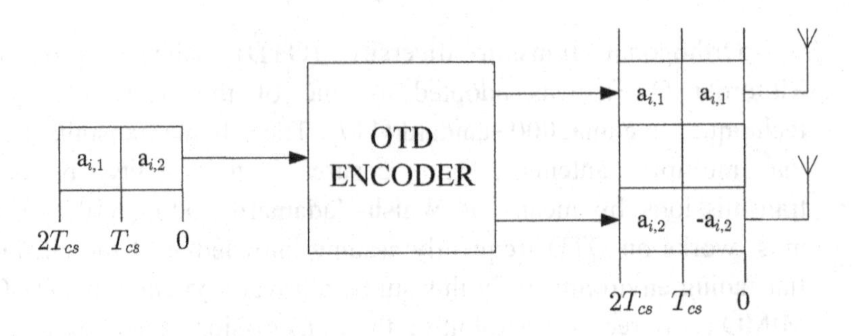

Figure 6.5: Orthogonal Transmit Diversity (OTD) encoding scheme.

necessary to ensure perfect orthogonality, that is $\underline{\alpha}_i^T \underline{\alpha}_j = 0$ for $\forall\, i \neq j$, among the spreading codes. Now since every symbol is repeated twice, the equivalent code signature vector attributed to the odd stream's antenna can be expressed as $\underline{\alpha}_{i1} = [\underline{\alpha}_i^T, \underline{\alpha}_i^T]^T$, and its complementary signature vector attributed to the even stream's antenna as $\underline{\alpha}_{i2} = [\underline{\alpha}_i^T, -\underline{\alpha}_i^T]^T$.

Suppose the transmitted signal from the jth antenna element (where $j \in \{1,2\}$) of the ith user arrives at the array receiver via K_{ij} multipaths, its associated channel impulse response corresponding to the kth path from the jth transmitting element of the ith user can thus be modelled as

$$\underline{\mathcal{I}}_{ijk}(t) \quad = \quad \underline{S}_{ijk}\beta_{ijk}\exp(j2\pi f_{ijk}t)\delta(t-\tau_{ijk}) \qquad (6.11)$$

where $\underline{S}_{ijk}$ denotes the receiver's *array-antenna* manifold vector, whilst τ_{ijk}, β_{ijk} and f_{ijk} are the path delay, complex fading coefficient and Doppler shift frequency respectively. Note that the formulation that follows can also be applied for the downlink by setting identical impulse response $\underline{\mathcal{I}}_{jk}(t)$ for all the M users, i.e. $\underline{\mathcal{I}}_{ijk}(t) = \underline{\mathcal{I}}_{jk}(t), \forall\, i$.

Taking the maximum delay spread to be within a symbol period, the received signal vector is oversampled at a rate of $1/T_s$ and passed through a bank of N_r tapped-delay lines (TDL), each of length $L = 3qN_c$ (with q being the chip oversampling factor). Upon concatenating the contents of the TDLs, the $N_r L$-dimensional discretised signal vector is thus formed and read for every two symbol periods. For completeness, its corresponding Doppler-STAR manifold vector is adapted and restated from previous study as follows

$$\underline{\mathfrak{h}}_{ijk}[n] = \underline{S}_{ijk} \otimes \left(\mathbb{J}^{l_{ijk}} \underline{\mathfrak{c}}_{ij} \odot \underline{\mathcal{F}}_{ijk}[n]\right) \tag{6.12}$$

where $l_{ijk} = \lceil \tau_{ijk}/T_s \rceil$ is the discretised delay, $\mathbb{J}$ is a $L \times L$ time down-shift operator, $\underline{\mathfrak{c}}_{ij}$ corresponds to the sampled version of the equivalent code signature associated with the jth antenna element of the ith user, i.e.

$$\underline{\mathfrak{c}}_{ij} = \sum_{p=0}^{2N_c-1} \alpha_{ij}[p] \,.\, \mathbb{J}^{pq} \underline{c} \tag{6.13}$$

with the vector $\underline{c}$ being the oversampled chip-level pulse shaping function $c(t)$ padded with zeros at the end,

$$\underline{c} = \left[c(0),\ c(T_s),\ \cdots,\ c((q-1)T_s),\ \underline{0}^T_{(L-q)}\right]^T \tag{6.14}$$

and $\underline{\mathcal{F}}_{ijk}[n]$ is the time-varying Doppler component modelled as

$$\underline{\mathcal{F}}_{ijk}[n] = \begin{bmatrix} 1 \\ \exp(j2\pi f_{ijk} T_s) \\ \exp(j2.2\pi f_{ijk} T_s) \\ \vdots \\ \exp(j2.(L-1)\pi f_{ijk} T_s) \end{bmatrix} \exp(j4n\pi f_{ijk} T_{cs}) \tag{6.15}$$

where, for simplicity in notation, the time index n is used to represent the every two-symbols periodical reading of the TDLs' contents.

Now by denoting the previous, current and next two consecutive repeated symbols as $\underline{\mathbb{a}}_{ij}[n] = \left[\, \mathbb{a}_{ij}[n-1],\, \mathbb{a}_{ij}[n],\, \mathbb{a}_{ij}[n+1]\,\right]^T$, the net discretised signal vector, incorporating Inter-Symbol Interference (ISI) and Multiple-Access Interference (MAI), can be written explicitly as

$$\underline{x}[n] \quad = \quad \sum_{i=1}^{M}\sum_{j=1}^{2} \bar{\mathbb{H}}_{ij}[n]\mathbb{G}_{ij}\underline{\mathbb{a}}_{ij}[n] + \underline{\mathbb{n}}[n] \tag{6.16}$$

where $\underline{\mathbb{n}}[n]$ is the sampled complex white Gaussian noise vector, $\mathbb{G}_{ij} = \mathbb{I}_3 \otimes \underline{\beta}_{ij}$ contains the multipaths' fading coefficients, and

$$\bar{\mathbb{H}}_{ij}[n] = \left[(\mathbb{I}_{N_r} \otimes (\mathbb{J}^T)^{2qN_c})\mathbb{H}_{ij}[n], \mathbb{H}_{ij}[n], (\mathbb{I}_{N_r} \otimes \mathbb{J}^{2qN_c})\mathbb{H}_{ij}[n]\right] \tag{6.17}$$

in which $\mathbb{H}_{ij}[n]$ has columns the Doppler-STAR manifold vectors of all the multipaths from the jth transmitting element of the ith user.

6.2.2 Blind Estimation and Reception

With the knowledge of the equivalent code signatures corresponding to the desired $\mathcal{D}^{\text{th}}$ user, a MUSIC-type channel estimation associated with each of the antenna elements can be performed by evaluating the cost function

$$\xi(\theta, l) \quad = \quad \frac{\left(\underline{S}(\theta) \otimes \mathbb{J}^l \underline{\mathbb{c}}_{\mathcal{D}j}\right)^H \mathbb{E}_{\mathrm{n}} \mathbb{E}_{\mathrm{n}}^H \left(\underline{S}(\theta) \otimes \mathbb{J}^l \underline{\mathbb{c}}_{\mathcal{D}j}\right)}{\left(\underline{S}(\theta) \otimes \mathbb{J}^l \underline{\mathbb{c}}_{\mathcal{D}j}\right)^H \left(\underline{S}(\theta) \otimes \mathbb{J}^l \underline{\mathbb{c}}_{\mathcal{D}j}\right)} \tag{6.18}$$

where $\underline{S}(\theta) \otimes \mathbb{J}^l \underline{\mathbb{c}}_{\mathcal{D}j}$ is the STAR manifold vector with $\underline{\mathbb{c}}_{\mathcal{D}j}$ being related to the equivalent code signatures due to the jth antenna element of the desired $\mathcal{D}^{\text{th}}$ user as shown in Equation (6.13), and $\mathbb{E}_{\mathrm{n}}$ is the matrix with columns the noise eigenvectors corresponding to the second order statistics of Equation (6.16).

Based on the estimated space-time channel parameters, the received discretised signal vector $\underline{x}[n]$ in Equation (6.16) is passed through a blind multipath filter bank as follows

$$\underline{y}[n] \quad = \quad \mathbb{F}^H . \underline{x}[n] \tag{6.19}$$

where $\mathbb{F} = \mathbb{P}_\mathrm{n}\widehat{\mathbb{B}}_\mathcal{D}\left(\widehat{\mathbb{B}}_\mathcal{D}^H\mathbb{P}_\mathrm{n}\widehat{\mathbb{B}}_\mathcal{D}\right)^{-1}$ with $\mathbb{P}_\mathrm{n}$ being the projection operator matrix onto the noise subspace of $\mathbb{R}_{xx} - \widehat{\mathbb{B}}_\mathcal{D}\widehat{\mathbb{B}}_\mathcal{D}^H$, in which $\mathbb{R}_{xx}$ is the covariance matrix of Equation (6.16), and $\widehat{\mathbb{B}}_\mathcal{D} = \left[\widehat{\mathbb{B}}_{\mathcal{D}1}, \widehat{\mathbb{B}}_{\mathcal{D}2}\right]$ contains columns the estimated STAR manifold vectors with the associated segmented matrix $\widehat{\mathbb{B}}_{\mathcal{D}j} \triangleq \left[\left(\underline{S}(\widehat{\theta}_{\mathcal{D}j1}) \otimes \mathbb{J}^{\widehat{l}_{\mathcal{D}j1}}\underline{\mathrm{c}}_{\mathcal{D}j}\right), \left(\underline{S}(\widehat{\theta}_{\mathcal{D}j2}) \otimes \mathbb{J}^{\widehat{l}_{\mathcal{D}j2}}\underline{\mathrm{c}}_{\mathcal{D}j}\right), \ldots, \left(\underline{S}(\widehat{\theta}_{\mathcal{D}jK_{\mathcal{D}j}}) \otimes \mathbb{J}^{\widehat{l}_{\mathcal{D}jK_{\mathcal{D}j}}}\underline{\mathrm{c}}_{\mathcal{D}j}\right)\right]$. Upon differential decoding, each of the segmented filter bank output is then equally gain combined to realise the nth symbol decision statistic for each of the transmitting antenna elements. Note that the small Doppler effect, instead of having to be estimated, has been mitigated by means of differential modulation.

6.2.3 Performance Analysis

Consider a uniform $N_r = 5$ element linear array (of half-wavelength spacing) at the receiver operating in the presence of $M = 3$ co-channel OTD users, where each user is assigned a unique PN-code sequence of length $\mathcal{N}_c = 31$. Let's take user 1 as the desired user, with an input SNR of 20dB, and the other 2 interferers each constitutes an interference ratio of 20dB (i.e. near-far problem). The chip rate is set at 1.2288Mchips/s with a carrier frequency of 2GHz; and its corresponding Doppler spread is set at 100Hz. Each user is assumed to have 8 multipaths and the array collects 200 data symbols for processing with chip-rate sampling of $q = 1$.

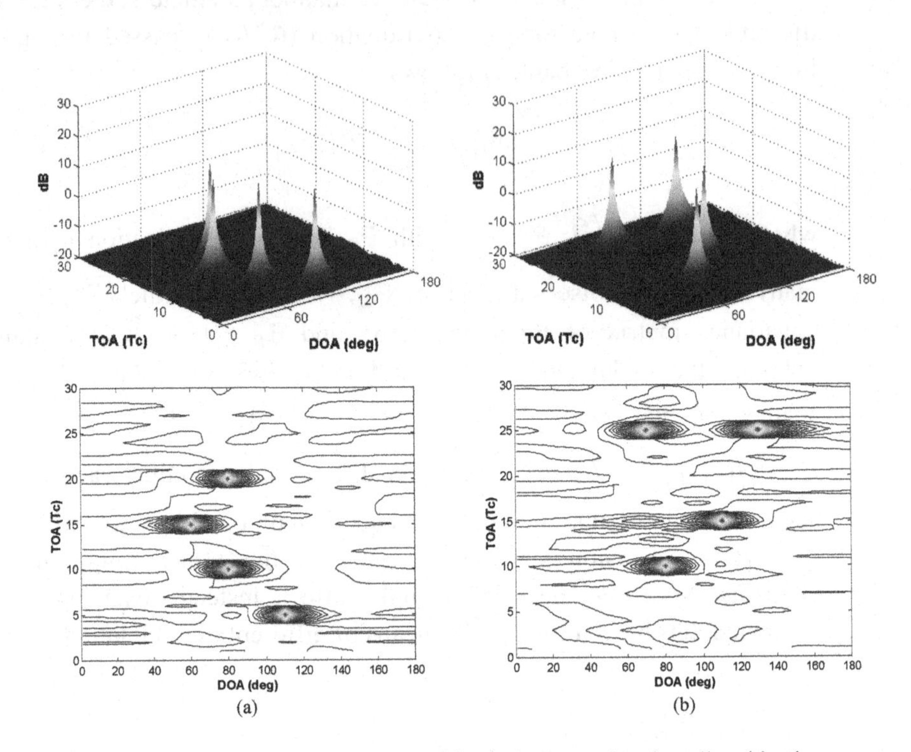

Figure 6.6: Space-time spectrum of the desired user showing all multipaths due to the (a) odd and (b) even stream's transmitting antennas.

The space-time estimation of all the 8 multipaths due to the desired user's odd and even stream's antennas is as illustrated in Figure 6.6. Notice that the algorithm can still operate even when the desired user's multipaths originated from its odd and even streams' antennas are co-located in both the space and time domains. Next, the performance of the proposed receiver is compared with a ST decorrelating detector and a conventional OTD receiver as depicted in Figure 6.7. It is therefore apparent that the BER curve due to the proposed receiver is significantly much lower, whilst the latter two deteriorate drastically even at low Doppler spread range.

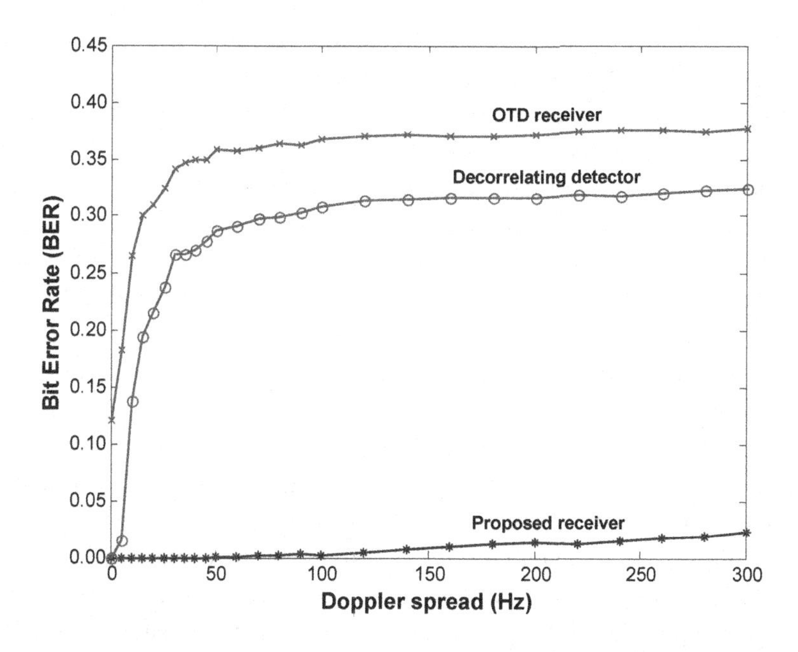

Figure 6.7: BER performance versus Doppler spread.

6.2.4 Concluding Summary

A blind near-far resistant MIMO array receiver, incorporating the concept of orthogonal transmit diversity, is proposed for asynchronous multipath DS-CDMA systems. With the integration of the *array-antenna* technology, the resulted receiver does not need any power control and it requires only the knowledge of the desired user's spreading code sequence.

6.3 MIMO Diversely Polarised Array System

Typical MIMO works frequently assume the use of multiple independent antenna elements, referred herein as *multiple-antenna* system, which are not sensitive to the polarisation of the incoming signals. To realistically model a practical MIMO channel, a diversely-polarised *array-antenna* reception is hence put forward in this study, in which a number of polarisation-sensitive antenna elements form an array system of a given geometry. The polarisation information inherent in the received signal, which is normally treated as polarisation fading, can be exploited to enhance the receiver's detection capability and improve its estimation accuracy and resolution. Furthermore, in addition to the spatial multiplexing technique commonly employed in most MIMO research, polarisation multiplexing can also be introduced to provide additional channel capacity. For instance, by utilising the three orthogonal polarisation states offered by a tripole sensor for polarisation multiplexing, the data rate of the system can be tripled effectively [118]. Not only that, additional domain of diversity in the form of polarisation diversity can now be attained, instead of the only spatial diversity being considered in existing MIMO system. Such diversity has been demonstrated in [119] where the same information symbol is transmitted via two orthogonal transverse fields, with each being spread by a distinct spreading code sequence. In view of all these advantages, this work thus proposed the use of diversely-polarised *array-antenna* reception in blind asynchronous MIMO system, implementing a combinatorial scheme which incorporates both the data-rate maximisation scheme and diversity maximisation scheme.

6.3.1 MIMO Diversely Polarised Signal Model

Let's consider an M asynchronous DS-CDMA users in a MIMO array system, with every user having N_t antenna elements, each capable of transmitting via P orthogonal polarisation states. The mode of transmission can be implemented in three ways. If $N_t = 1$, the scheme can take on either the data rate maximisation scheme (scheme 1) or diversity maximisation scheme (scheme 2). The former scheme is performed by demultiplexing a high bit-rate input signal source into P data streams, and send simultaneously using different polarisation states. The latter scheme, on the other hand, sends P copies of the same input signal source over each of the polarisation states to maximise the diversity advantages in fading channels. However if $N_t \neq 1$, a combinatorial scheme (scheme 3) can be employed - the high bit-rate input signal source is first duplicated into N_t data streams, with each later demultiplexed into P substreams and transmitted via the different polarisation states. In all the above three schemes, a unique spreading code of length $\mathcal{N}_c$ is assigned to each user to be applied across its transmitting elements. Suppose the signal transmitted with the oth polarisation state from the jth antenna element of the ith user arrives at the receiver, employing an array of N_r polarisation-sensitive sensors (e.g. crossed-dipole or circularly-polarised sensors), via K_{ijo} multipaths. By denoting its previous, current and next differential encoded data symbol as $\underline{a}_{io}[n] = \left[\mathrm{a}_{io}[n-1],\, \mathrm{a}_{io}[n],\, \mathrm{a}_{io}[n+1]\right]^T$, the discretised received signal vector, incorporating Inter-Symbol Interference (ISI), Multiple-Access Interference (MAI), and co-code interferences, can hence be expressed explicitly as follows

$$\underline{x}[n] \quad = \quad \sum_{i=1}^{M}\sum_{j=1}^{N_t}\sum_{o=1}^{P}\overline{\mathbb{H}}_{ijo}\mathbb{G}_{ijo}\underline{a}_{io}[n] + \underline{n}[n] \tag{6.20}$$

where $\underline{n}[n]$ is the sampled complex white Gaussian noise vector, $\mathbb{G}_{ijo} = \mathbb{I}_3 \otimes \underline{\beta}_{ijo}$ with $\underline{\beta}_{ijo}$ the complex fading coefficient, and $\overline{\mathbb{H}}_{ijo} = \left[\, (\mathbb{I}_{N_r} \otimes (\mathbb{J}^T)^{L/2})\mathbb{H}_{ijo},\, \mathbb{H}_{ijo},\, (\mathbb{I}_{N_r} \otimes \mathbb{J}^{L/2})\mathbb{H}_{ijo} \,\right]$, in which $\mathbb{J}$ (or $\mathbb{J}^T$) is a $L \times L$ time down-shift (or up-shift) matrix, and $\mathbb{H}_{ijo} = \left[\underline{\mathfrak{h}}_{ijo1}, \underline{\mathfrak{h}}_{ijo2}, \ldots, \underline{\mathfrak{h}}_{ijoK_{ijo}}\right]$ with $\underline{\mathfrak{h}}_{ijok}$ being the Polar-STAR manifold vector associated with the oth polarisation state, encompassing the path's delay, arriving angle and state of polarisation, as similar to that in Equation (3.17).

6.3.2 Blind Estimation and Reception

By applying the PADE algorithmic routine detailed from Section 3.4.1 to Section 3.4.3, the Polar-STAR manifold vector of the ith user can be reconstructed based on the estimated polarisation-space-time channel parameters to obtain $\widehat{\mathbb{H}}_i = [\widehat{\mathbb{H}}_{i1}, \widehat{\mathbb{H}}_{i2}, \ldots, \widehat{\mathbb{H}}_{iN_t}]$, where $\widehat{\mathbb{H}}_{ij} = [\widehat{\mathbb{H}}_{ij1}, \widehat{\mathbb{H}}_{ij2}, \ldots, \widehat{\mathbb{H}}_{ijP}]\ \forall j = 1, 2, \ldots, N_t$.

To suppress the inherent interferences, the received discretised signal vector $\underline{x}[n]$ in Equation (6.20) is passed through a novel polarimetric-spatial-temporal multipath filter bank based on the orthogonal projection of the interference subspace to obtain

$$\underline{y}[n] = \mathbb{L}^H . \underline{x}[n] \tag{6.21}$$

where the matrix $\mathbb{L} = \mathbb{P}^{\perp}_{\widehat{\mathbb{H}}_{intf}} \widehat{\mathbb{H}}_{\mathcal{D}} \left(\widehat{\mathbb{H}}_{\mathcal{D}}^H \mathbb{P}^{\perp}_{\widehat{\mathbb{H}}_{intf}} \widehat{\mathbb{H}}_{\mathcal{D}}\right)^{-1}$ with $\widehat{\mathbb{H}}_{intf} = \left[(\mathbb{I}_{N_r} \otimes (\mathbb{J}^T)^{L/2})\widehat{\mathbb{H}}_{com}, \widehat{\mathbb{H}}^{(\mathcal{D})}_{com}, (\mathbb{I}_{N_r} \otimes \mathbb{J}^{L/2})\widehat{\mathbb{H}}_{com} \right]$ in which $\widehat{\mathbb{H}}_{com} = \left[\widehat{\mathbb{H}}_1, \widehat{\mathbb{H}}_2, \ldots, \widehat{\mathbb{H}}_M \right]$ is the estimated composite channel parameters and $\widehat{\mathbb{H}}^{(\mathcal{D})}_{com}$ is the composite matrix $\widehat{\mathbb{H}}_{com}$ with the exclusion of the desired $\mathcal{D}^{\text{th}}$ user's matrix $\widehat{\mathbb{H}}_{\mathcal{D}}$.

If scheme 2 is employed in the system, the filter bank output can be simply combined to realise the nth symbol decision statistic, i.e.

$$b[n] = \underline{w}^H . \underline{y}[n] \tag{6.22}$$

where $\underline{w}$ is the combining weight vector obtained using the principal eigenvector of the autocorrelation matrix of Equation (6.21).

However, for the other two schemes, the estimated signal path needs to be segregated according to its polarisation state. Taking each state $o \in \{1, 2, \ldots, P\}$ as the reference point (γ_o, η_o), the polarimetric distribution of the detected multipaths can be evaluated as

$$\begin{aligned}\delta_{\mathcal{D}jko} = \arccos\big[& \cos(2\gamma_o)\cos(2\widehat{\gamma}_{\mathcal{D}jko}) + \\ & \sin(2\gamma_o)\sin(2\widehat{\gamma}_{\mathcal{D}jko})\cos(\eta_o - \widehat{\eta}_{\mathcal{D}jko})\big]\end{aligned} \tag{6.23}$$

where $0 \leq \delta_{\mathcal{D}jko} \leq \pi$ is the path angular distance with respect to the reference point (γ_o, η_o) on the Poincaré sphere. If the multipaths' polarimetric distances fall within a prespecified polarimetric decision region (i.e. $\delta_{\mathcal{D}jko} \leq \zeta_o$), they are grouped together under the oth polarisation state. This can then be reflected in the construction of the filter by arranging the Polar-STAR manifold vectors within the matrix $\widehat{\mathbb{H}}_{\mathcal{D}}$ in a segmented manner with each segment manifesting each polarisation state. Each of the P segments of the filter bank outputs can then be combined using Equation (6.22) and subsequently multiplexed together to realise the high bit rate data stream.

6.3.3 Performance Analysis

Consider a uniform linear $N_r = 5$ elements diversely-polarised array (of half-wavelength spacing) with $M = 3$ co-channel users utilising a combinatorial transmission scheme. Each user is assigned a unique Gold sequence of length $\mathcal{N}_c = 31$ to be shared across $N_t = 2$ antenna elements with each capable of transmitting $P = 2$ orthogonal polarisation states: Left-Hand Circular (LHC) and Right-Hand Circular (RHC). Let's take user 1 as the desired user, with an input SNR of 10dB, while the other 2 interferers each constituting an interference ratio of 10dB (i.e. near-far problem). All 3 users have 8 multipaths each and the array collects 150 data symbols, with a chip-rate sampler, for processing.

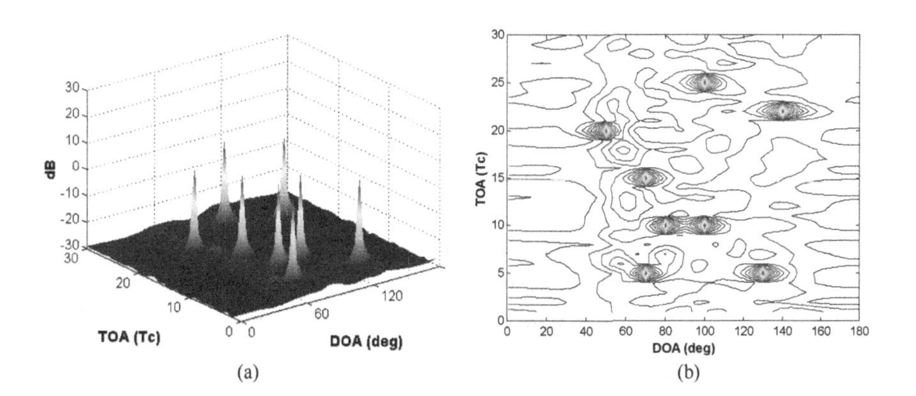

Figure 6.8: Space-time spectrum of all the polarimetric multipaths associated with the desired user 1.

The channel parameters of all the 8 multipaths due to the desired user are estimated by applying the PADE algorithm to obtain the space-time spectrum plot as depicted in Figure 6.8. Now since there are only 2 polarisation states, the prespecified polarimetric region can be set at $\zeta = \pi/2$ with respect to the LHC point so that the upper hemisphere of the Poincaré sphere is attributed to the LHC state, while the lower is attributed to the RHC state as illustrated in Figure 6.9.

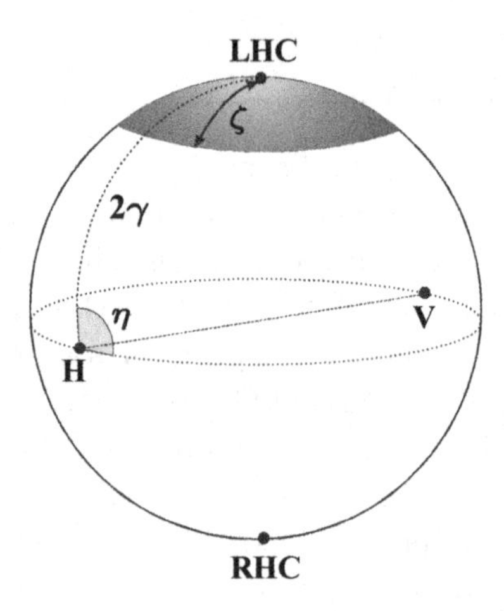

Figure 6.9: Prespecified polarimetric region ζ on Poincaré sphere.

Table 6.1 shows the polarimetric distribution of all the estimated multipaths. The proposed receiver's signal constellation is then compared with its equivalent space-time-polar decorrelating detector and 3D RAKE receiver, with the latter two assuming full knowledge of the channel. It is thus evident from Figure 6.10 that the proposed blind receiver performs just as well as the decorrelating detector, while the RAKE receiver is overwhelmed with interferences.

Table 6.1: Multipath polarimetric distribution of desired user $i = 1$								
Path (j,k,o)	(1,1,1)	(1,2,1)	(1,3,1)	(1,1,2)	(1,2,2)	(2,1,1)	(2,1,2)	(2,2,2)
δ_{1jko}	13.9°	21.5°	0.8°	143.4°	151.7°	27.8°	165.3°	177.6°

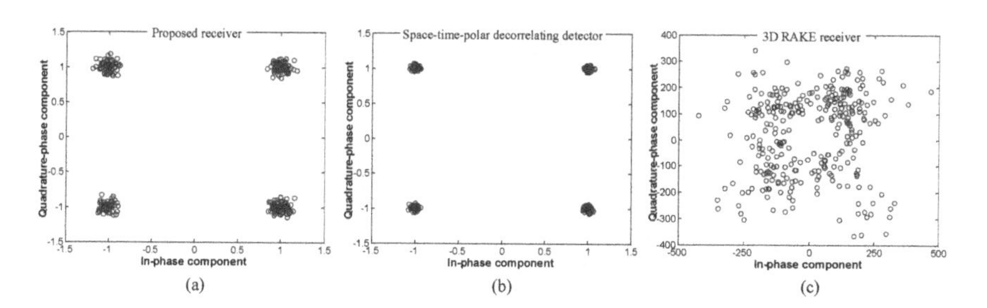

Figure 6.10: Signal constellation associated with the (a) proposed receiver, (b) space-time-polar decorrelating detector and (c) 3D RAKE receiver.

6.3.4 Concluding Summary

In this work, a blind near-far resistant MIMO array receiver employing diversely polarised array is proposed for asynchronous multipath DS-CDMA systems. By means of polarisation multiplexing, the proposed MIMO system is applicable in either (i) the data rate maximisation scheme, (ii) the diversity maximisation scheme, or (iii) the combinatorial maximisation scheme, without exhausting the limited number of spreading codes available.

CHAPTER 7

Discussions on Present and Future Research Work

7.1 Key Points of Current Work

The proliferation of portable mobile devices has called forth the development of the future enabling technology beyond the existing wireless communications network. The two emerging technological fields: space-time array processing and code division multiple access, hence are envisaged to be the principal proponents in the advancement and the sustainment of the next-generation mobile telecommunications. In view of the multipath channel propagation impairments, the integration of the two technologies aids not only in complementing one another but also in synergising both of their advantages. In fact, by employing sensor array processing in CDMA system, additional dimensions of interference cancellation and new countermeasures in handling undesirable channel effects can thus be devised.

This therefore establishes the driver behind the Spatial-Temporal ARray (STAR) architecture which lays the basic framework for the works proposed in the study. There are essentially three parts in the framework. First is the spread-spectrum multiple access transmitter which herein does not impose the assumptions of any power control (i.e. near-far effect) or any synchronism (i.e. asynchronous transmittal). Second is the space-time array receiver where the spatial-temporal array manifold vector is introduced based on an extension from the well-known spatial array manifold vector. Third is the multiuser vector channel that provides the propagation link between both ends of the transmitter and receiver. The vector channel model is a powerful tool in mapping the transmitted signal at the transmitter, after passing through the channel, to the received signal at the receiver. It renders the spatial and temporal characterisation of the multipath propagation environment between which the signals are transmitted into and received from the channel. These three primary parts constitute the overall STAR architectural framework which put forward the groundwork in the mathematical modelling for the subsequent system formulations.

By incorporating the polarisation information inherent in the signal, the architectural framework is extended to that based on the Polar-STAR manifold vector. The motivation behind it is due to the fact that in typical wireless communications systems, multiple copies of the signal are created with different arriving angle, path delay, polarisation and fading, as a result of its complex interactions with the propagation channel. But despite the existence of this diverse information inherent in the multipath signals, its polarisation aspect is often being overlooked as an additional means of signal discrimination. In fact, most array processing techniques often assume the use of polarisation-insensitive sensors, disregarding the reality whereby many of the practical arrays are sensitive to the polarisation of the received signals. Furthermore, the received signal polarisation can be substantially diverse, considering the recent development of portable handheld terminals which are often randomly orientated, and the depolarising nature of the propagation channel. This diverse polarisation factor, instead of being considered as part of signal fading (i.e. polarisation fading), can be exploited to enhance the receiver's detection capability and its estimation accuracy/

resolution. This can be achieved by means of a diversely polarised array which has been studied in a number of direction finding algorithms. In order to alleviate the multidimensional operation into its polarisation search space, an efficient near-far resistant polarisation-angle-delay estimation (PADE) algorithm is proposed which provides a joint space-time estimate of the desired user in an asynchronous multipath environment. Its associated polarisation parameters, if required, can also be obtained using a set of analytical equations. Due to the inclusion of the polarisation dimension via the use of the diversely-polarised array, the joint space-time channel estimation is now able to resolve closely-located paths that otherwise cannot be resolved using its equivalent polarisation-insensitive array. In the instances whereby the paths are so closely located such that the proposed cost function fails in their spatial and temporal differentiation, its supplementary cost function can be employed to detect such occurrences.

Having established the formulation of the PADE algorithm, the approach is then applied to the Diffuse-STAR framework. A simplistic view of the multipath environment, making up of multiple point sources, is frequently being adopted in traditional array processing model. But in an urban or suburban setup, diffusion in the signal component is inevitable as a result of the scattering mechanism encountered by the propagating signal in the channel. The diffuse signal is composed of a superposition of multiple point sources that creates the spreading phenomenon as seen in the distributed cluster. This thus results in performance degradation when conventional processing techniques based on point sources assumption are applied. A generalised diffusion framework is therefore proposed in the system design so as to handle the occurrences of multiple point and/or diffuse sources resulted in the signal environment. By adapting the previously derived PADE algorithm in the diffusion case, its spatial and temporal channel parameters associated with the diffuse signal can be jointly estimated in conjunction with its corresponding weighted perturbation factor. Likewise the channel estimator, integrated at the front-end of the proposed blind diffusion-based receiver, is insusceptible to the near-far problem effect. Furthermore, its underlying architecture provides ease in the extension to a co-code environment (brought about by co-code interferers in for instance a code-reuse scenario or a jamming situation) by simply introducing a novel *Branch Identification Process* in the detection procedure. With that, the receiver is robust to any instances of erroneous (incorrect channel estimation) or unidentified (incomplete channel estimation) channel parameter resulted at the estimation phase; and its reception process only requires the availability of a single estimable diffuse signal due to the desired user that is differentiable in at least one of the space, time and code domains.

Given the generalised diffusion-based receiver operating in the presence of co-code interferers, parallels can be drawn from its architectural framework to that of the Multiple Input and Multiple Output (MIMO) system. Unlike many reported works on MIMO which are based on multiple independent antenna elements, the proposed MIMO array receiver is developed based on the *array-antenna* technology which is commonly used in applications such as the well-known smart-antenna system. By harnessing the attributes of the *array-antenna*, an extra layer of co-channel interference cancellation and new ways for handling unwanted channel effects can be developed. The advantage of such incorporation has been demonstrated by a novel Doppler-STAR MIMO array receiver proposed for time-varying asynchronous multipath DS-CDMA system. With that, its Doppler-space-time channel estimation algorithm can be devised for the joint angle and delay, as well as the Doppler shift frequency estimation. The multipath's Doppler effect, instead of being detrimental to the system performance, is been exploited in the receiver design to provide an additional domain of diversity. Furthermore, the proposed MIMO array receiver does not require the imposition of the knowledge of the channel or the need of any power control, as are readily assumed in conventional MIMO systems. It is also applicable in the two MIMO transmission schemes, that is the data rate maximisation scheme or the diversity maximisation scheme, without having to exhaust the limited number of available spreading codes. In addition, the spreading code assignment strategies employed in typical CDMA BLAST system can also be adapted in the proposed formulation.

To illustrate the potential benefit of adopting the STAR architectural framework, the investigation is extended to include other application examples to gain a further insight and new perspective in the estimation and reception process. Three examples are proposed and presented herein. One is by utilising the joint spatial and temporal dimensions to discern each of the multipaths in order to facilitate the subsequent Doppler estimation process. The other is by exploiting the orthogonal transmit diversity technique to aid in the means of segregating the multipaths, identified from the channel estimation phase, to each of the transmitting antenna elements. And the last is by applying polarisation multiplexing so that each of the multipaths can be discriminated by its associated inherent state of polarisation. Likewise, all the three proposed estimation and reception algorithms possess the capabilities of being blind and near-far resistant.

7.2 Directions for Further Research

Electromagnetic Vector-Sensor Array Communications

Electromagnetic vector-sensor is a term coined for a sensor element with its vector output generating the separate measurement of the complete electric and magnetic field components as a result of an incident TEM signal wavefield. Unlike the diversely polarised sensors considered thus far, the vector-sensor is capable of taking advantage all the inherent electromagnetic field information put forward by the propagating signal. These sensors are already available commercially, for example, from Flam and Russell Inc. in Horsham, PA, and from EMC Baden Ltd. in Baden, Switzerland. Since then, Lincoln Lab at the Massachusetts Institute of Technology, Cambridge has performed some preliminary localisation tests utilising the Compact Array Radiolocation Technology (CART) array manufactured by the Flam and Russell Inc. [120]. The array is composed of three orthogonally orientated short dipoles and three orthogonally orientated magnetic loops, all collocated in a point-like geometrical space [121]. Nehorai and Paldi has also pioneered a simple but novel idea of using the vector cross-product of the electric field vector output and the magnetic field vector output, provided by the employment of only a single vector-sensor, to directly estimate both the azimuth and elevation radial directions of the targeted source [122]. This constitutes a significant property of the vector-sensor in which its direction finding capability in the 3D space can be realised with just solely a single sensor, in contrast with the conventional requirement of a 2D planar sensor array that normally demands a sizeable space imposition. Thence it will be interesting, especially in applications whereby space is a constraint, to extend the space-time estimation and reception algorithm, by means of a single or an array of electromagnetic vector-sensor(s), to incorporate the full exploitation of both the electric and magnetic field aspects of the multipath signals.

Applications of PADE Algorithm

The PADE algorithm, developed in Section 3.4.3, is not specifically derived for the polarisation-angle-delay estimation purposes only. The algorithmic concept can also be applied to other applications as illustrated in the generalised diffusion framework for point and/or diffused sources as described in Section 4.4.1. Both cases require a multivariate objective function in its estimation phase which necessitates an exhaustive search in a multidimensional search space. But with the exploitation of the PADE concept, this is consequently simplified to a 2D cost function yielding the joint estimation of the space-time channel parameters in conjunction with a set of analytical expressions providing its associated accompanying factor(s). The generality in the application of the PADE algorithm is therefore evidential. In fact, a similar approach has also been adapted to a direction finding application utilising partly-calibrated sensor array with an arbitrary geometry [123]. However this time, its multidimensional search function is simplified to a 1D spectral search in the space domain, together with an unknown factor capturing its corresponding uncalibrated manifold component. Note that although only the azimuth angle is used in its 1D search space, the cost function can be easily extended to a 2D direction finding problem encompassing both the azimuth and elevation directional angles, or to a 2D joint space-time channel estimation problem operating in the presence of partial array calibration.

MC-CDMA Based MIMO Array System

Multi-Carrier Code Division Multiple Access (MC-CDMA) is a powerful and forthcoming technique in the context of high bit-rate services for the future wireless mobile communications system. Having seen the various configurations of the MIMO array system, it is henceforth worthwhile extending the configurational concept to the MC-CDMA modulation scheme. The scheme, which is a combination of OFDM and CDMA, embodies significant property that is desirable for implementation in the MIMO system. By splitting the high-speed data input into multiple low-rate streams transmitted simultaneously via a set of subcarriers, the MC-CDMA scheme has essentially transformed the corresponding frequency-selective channel into a set of parallel narrowband frequency-flat fading subchannels. This flat-fading characteristic at each subcarrier is ideally suited to the application of the Space-Time Block Coding (STBC) strategy which is originally meant for the frequency non-selective MIMO environment when first proposed by Alamouti [102]. Preliminary implementation of the space-time coding strategy, such as the Space-Time Transmit Diversity (STTD) and Space-Time Spreading (STS) schemes, have been introduced in blind near-far resistant MIMO CDMA array systems [124]. Parallels may be drawn to extend the framework to that of the MC-CDMA modulation scheme so as to fully exploit the frequency-flat STBC mechanisms [125].

Collaborative Processing for Wireless Sensor Network

Wireless sensor network [126] is made up of a number of sensor nodes which can be deployed in a wide variety of applications. The rationale behind the distributed sensor network is purposed for detecting, identifying, localising, monitoring, or tracking one or more targets of interest. These networks may be used in healthcare telemetry, smart-home network, military or security surveillance, natural-catastrophe monitoring, industrial control and automation, consumer electronics, and many others. However, active research challenge remains in the design of the processing algorithms that are robust against demanding environmental conditions and strong interference occurrences. These sensor nodes may be operating in an uncertain channel environment, for instance on a rough terrain with intentional jammers' disruption, thence consequently compromising the reliability of the system. Sensor array processing, which has been seen to exploit the inherent diversity in the environment and provide new ways of interference suppression, is therefore envisaged to be capable in enhancing the wireless transmission links of the sensor networks. The array can be formed by clustering two or more sensor nodes, with each consisting of an antenna-element or an antenna-subarray, for inter-node and/or intra-node collaborative array processing [127]. In addition to that, another interesting idea for further research would be to look at this multi-node clustering approach under the MIMO framework. Its multi-transmission schemes, designed for diversity maximisation or data-rate maximisation, can be extended to the cluster which is a desirable feature for the collaborative sensor network.

Positioning in Cellular Geolocation System

Cellular geolocation renders localisation estimates of the mobile terminals as they transmit over existing cellular base stations infrastructure. It enables the positioning of any portable handheld devices in the cellular network without the need of having a GPS receiver. This consequently transfers the computational load from the mobile terminal to the base station, hence prolonging the battery lifespan of the portable handheld devices. More importantly, cellular geolocation enables the determination of the system capacity needs in any particular region so as to aid in the appropriate adaptation of its network resources accordingly. The positioning system is generally implemented in three ways. First is the DOA method (also known as triangulation) whereby the location of the target is found by the intersection of two or more Lines Of Bearing (LOBs) provided by the base stations. Second is the TOA method in which its source location is obtained by the intersection of two or more Time Difference of Arrival (TDOA) hyperboloids generated from the base stations. Third is the hybrid combination of both the techniques as described in [128]. Incidentally in this study, the near-far resistant space-time estimation algorithms established in the three different environmental scenarios are of value in their own merits for the location positioning applications. Further investigation of this joint DOA and TOA estimation algorithm in the design of the cellular geolocation methodology could therefore prove promising.

Abbreviations & Notations

I. Mathematical Conventions

A	Scalar
$\underline{A}$	Vector
$\mathbb{A}$	Matrix
$\underline{A}^b$	Element by element power
$\mathbb{I}_N$	Identity matrix of $N \times N$ dimension
$\underline{0}_N$	Zero vector of N elements
$\mathbb{P}_{\mathbb{A}}$	Projection matrix onto the subspace of matrix $\mathbb{A}$
$\mathbb{P}_{\mathbb{A}}^{\perp}$	Orthogonal projection matrix onto the subspace of matrix $\mathbb{A}$
$\arg(A)$	Argument of A
$\det(\mathbb{A})$	Determinant of matrix $\mathbb{A}$
$\mathrm{diag}(\underline{A})$	Diagonalisation of vector $\underline{A}$
$\exp(\underline{A})$	Element by element exponential of vector $\underline{A}$
$\mathrm{trace}(\mathbb{A})$	Trace of matrix $\mathbb{A}$
$\mathrm{vec}(\mathbb{A})$	Row-wise vectorisation of matrix $\mathbb{A}$
$(\bullet)^H$	Hermitian transpose
$(\bullet)^T$	Transpose
$(\bullet)^*$	Complex conjugate
$\lceil \bullet \rceil$	Round up to integer
$\lvert \bullet \rvert$	Absolute value
$\lVert \bullet \rVert$	Euclidean norm
$\otimes$	Kronecker product
$\odot$	Hadamard (Schur) product
$\oslash$	Hadamard (Elementwise) division
$\mathcal{R}$	Set of real numbers
$\mathcal{C}$	Set of complex numbers
$\mathcal{N}$	Set of natural numbers
$\mathcal{Z}$	Set of integers

II. Symbols

$\mathrm{a}[n]$	Data symbol
$b[n]$	Decision statistic of nth symbol
$c(t)$	Chip pulse shaping waveform
$c_{PN}(t)$	Pseudo-noise spreading waveform
$\delta(t)$	Dirac's delta function
$\mathcal{F}(t)$	Doppler shift component
$\underline{\mathcal{F}}[n]$	Sampled Doppler shift component
$\underline{\mathcal{I}}(t)$	Vector channel impulse response
$\underline{\mathcal{I}}[n]$	Discrete channel response vector
$\underline{I}_{\text{ISI}}[n]$	Inter-Symbol Interference component
$\underline{I}_{\text{MAI}}[n]$	Multiple-Access Interference component
$m(t)$	Complex baseband DS-CDMA signal
$\underline{\mathrm{n}}(t)$	Complex noise vector
$\underline{\mathrm{n}}[n]$	Sampled noise vector
$\underline{\mathrm{s}}(t)$	Complex baseband signal of scattering cluster
$\underline{x}(t)$	Continuous-time received baseband signal
$\underline{x}[n]$	Discrete-time stacked received data vector
$\mathbb{A}$	Matrix with columns containing the modified manifold vector $\underline{A}$
$\underline{A}$	Modified array manifold vector
α	Spreading code chip
$\underline{\alpha}$	Spreading code sequence
B	Number of unique code signatures
β	Complex path coefficient
c	Speed of light
$\underline{\mathrm{c}}$	Discretised spreading code vector
$\underline{\tilde{\mathrm{c}}}$	Fourier transformed version of $\underline{\mathrm{c}}$
$\mathcal{D}$	Desired user of interest
δ	Angular distance on Poincaré sphere

$\mathbb{E}_n$	Matrix with columns containing the noise eigenvectors
$\underline{E}$	Electric field of TEM wave
E_ϕ, E_θ	Horizontal and vertical electric field components
E_x, E_y, E_z	Electric fields in the x, y and z directions
$\mathbb{F}$	Fourier transformation matrix
F_c	Signal carrier frequency
f	Doppler shift frequency
γ, η	Polarisation parameters
Υ	Intra-symbol Doppler constituent
Ψ	Inter-symbol Doppler constituent
$\mathbb{H}$	Matrix with columns containing the manifold vector $\underline{\mathfrak{h}}$
$\underline{\mathfrak{h}}$	Polar/Diffuse/Doppler-STAR manifold vector
$\mathbb{J}, \mathbb{J}^T$	Time down-shift and up-shift operator matrix
K	Number of multipath rays
$\underline{k}(\theta, \phi)$	Wavenumber vector
κ	Aggregate fading coefficient of scattering cluster
$\mathbb{L}$	Filter bank
L	Length of tapped-delay line
l	Discretised path delay
λ	Signal wavelength
$\mathbb{M}$	Matrix constituent of modified array manifold vector
M	Number of users
N	Number of sensor elements in antenna array
N_r	Number of receiving antenna elements
N_t	Number of transmitting antenna elements
$\mathcal{N}_c$	Number of chips per symbol
$\mathcal{O}$	Number of filter bank branches
$\underline{p}$	Vector constituent of modified array manifold vector
q	Oversampling factor

$\underline{q}$	Induced electric field components
ϱ	Fractional delay factor
$\mathbb{R}$	Covariance matrix
$\underline{r}_x, \underline{r}_y, \underline{r}_z$	Cartesian coordinates of antenna array geometry
$\mathbb{S}$	Matrix with columns containing the spatial manifold vector
$\underline{S}$	Spatial array manifold vector
$\underline{\dot{S}}$	First order derivative of spatial array manifold vector
σ	Angular spread of scattering cluster
$\mathbb{T}(\theta, \phi)$	Spherical-to-Cartesian transformation matrix
T_c	Chip period
T_s	Sampling period
T_{cs}	Channel symbol period
τ	Signal path delay
$\underline{\Theta}$	Path's propagation state
$\widetilde{\theta}$	Angular perturbation about cluster's nominal direction
θ, ϕ	Direction angle at azimuth and elevation planes
U	Number of co-code users
$\underline{u}$	Unit vector
$\mathbb{V}$	Sensor's sensitivities of antenna array
$\underline{\mathcal{V}}$	Sensor's sensitivities in the x, y, and z directions
V	Induced voltage by unit electric field
v	Velocity of motion towards receiver
φ	Weighted perturbation factor of scattering cluster
ϖ	Axial ratio of ellipse locus
W	Number of scatterers in scattering cluster
$\underline{w}$	Weight vector
ξ	Cost function
$\xi_{\min}$	Cost function minima value
$\mathbb{Z}$	Preprocessor operator matrix

III. Acronyms

ADC	Analogue-to-Digital Conversion
AMPS	Advanced Mobile Phone Service
AWGN	Additive White Gaussian Noise
BER	Bit-Error Rate
BLAST	Bell Labs LAyered Space-Time
BPSK	Binary Phase Shift Keying
CART	Compact Array Radiolocation Technology
CCI	Co-Channel Interference
CDMA	Code Division Multiple Access
CML	Conditional Maximum Likelihood
COLD	Co-centered Orthogonal Loop and Dipole
D-AMPS	Digital Advanced Mobile Phone Service
Diffuse-STAR	Diffuse-Spatial-Temporal ARray
DISPARE	DIspersed Signal PARametric Estimation
DML	Deterministic Maximum Likelihood
DOA	Direction-of-Arrival
Doppler-STAR	Doppler-Spatial-Temporal ARray
DPSE	Distributed Signal Parameter Estimator
DS-CDMA	Direct-Sequence Code Division Multiple Access
ESPRIT	Est. of Sig. Parameters via Rotational Invariance Technique
ETSI	European Telecommunications Standards Institute
FDMA	Frequency Division Multiple Access
FH-CDMA	Frequency-Hopping Code Division Multiple Access
FPLMTS	Future Public Land Mobile Telephone Systems
GPS	Global Positioning System
GSM	Global System for Mobile communications
IMT-2000	International Mobile Telecommunications 2000
IS-54	Interim Standard for U.S. Digital Cellular
IS-95	Interim Standard for U.S. Code Division Multiple Access
ISI	Inter-Symbol Interference
ITU	International Telecommunication Union
JTACS	Japanese Total Access Communications System
LHC	Left-Hand Circular
LOB	Line Of Bearing
LOS	Line-Of-Sight
MAI	Multiple Access Interference
MC-CDMA	Multi-Carrier Code Division Multiple Access
MIMO	Multiple Input and Multiple Output
Min-Norm	Minimum-Norm
MUSIC	MUltiple SIgnal Classification
MVDR	Minimum Variance Distortionless Response
NFR	Near-Far Ratio

NLOS	Non Line-Of-Sight
NMT	Nordic Mobile Telephone
NTT	Nippon Telegraph and Telephone
OFDM	Orthogonal Frequency Division Multiplexing
OTD	Orthogonal Transmit Diversity
PADE	Polarisation-Angle-Delay Estimation
PDC	Personal Digital Cellular
PN	Pseudo-Noise
Polar-STAR	Polarisation-Spatial-Temporal ARray
PSTN	Public Switched Telephone Network
QoS	Quality of Service
QPSK	Quadrature Phase Shift Keying
RHC	Right-Hand Circular
RMSE	Root Mean Square Error
SDMA	Space Division Multiple Access
SIMO	Single Input and Multiple Output
SIR	Signal-to-Interference Ratio
SIVO	Scalar-Input Vector-Output
SML	Stochastic Maximum Likelihood
SNIR	Signal-to-Noise-and-Interference Ratio
SNR	Signal-to-Noise Ratio
SPD	Spread
SS	Spread Spectrum
ST	Space-Time
STAR	Spatial-Temporal ARray
STBC	Space-Time Block Coding
STS	Space-Time Spreading
STTD	Space-Time Transmit Diversity
TACS	Total Access Communications System
TDL	Tapped-Delay Line
TDMA	Time Division Multiple Access
TDOA	Time Difference of Arrival
TEM	Transverse ElectroMagnetic
TH-CDMA	Time-Hopping Code Division Multiple Access
TOA	Time-of-Arrival
ULA	Uniform Linear Array
UML	Unconditional Maximum Likelihood
UMTS	Universal Mobile Telecommunications System
VIVO	Vector-Input Vector-Output
WCDMA	Wideband Code Division Multiple Access
WSF	Weighted Subspace Fitting
1D/2D/3D	One-Dimensional/Two-Dimensional/Three-Dimensional
1G/2G/3G	First Generation/Second Generation/Third Generation

Bibliography

[1] F. Nack, "Migrating from mobile telephony to multipurpose gadgets", *IEEE Multimedia*, vol. 10, no. 2, pp. 8-11, Apr-Jun 2003.

[2] J. E. Padgett, C. G. Gunther and T. Hattori, "Overview of wireless personal communications", *IEEE Communications Magazine*, vol. 33, no. 1, pp. 28-41, Jan 1995.

[3] D. J. Goodman, "Second generation wireless information networks", *IEEE Trans. Vehicular Technology*, vol. 40, no. 2, pp. 366-374, May 1991.

[4] P. Sehier, J. M. Gabriagues, and A. Urie, "Standardization of 3G Mobile Systems", Alcatel Telecommunications Review - White Papers, pp. 10-18, Mar 2001.

[5] T. Ojanpera and R. Prasad, "An overview of third-generation wireless personal communications: a European perspective", *IEEE Personal Communications*, vol. 5, no. 6, pp. 59-65, Dec 1998.

[6] Y. Kim, B. J. Jeong, J. Chung, C. Hwang, J. S. Ryu, K. Kim and Y. K. Kim, "Beyond 3G: vision, requirements, and enabling technologies", *IEEE Communications Magazine*, vol. 41, no. 3, pp. 120-124, Mar 2003.

[7] J. S. Thompson, P. M. Grant and B. Mulgrew, "Smart antenna arrays for CDMA systems", *IEEE Personal Communications*, vol. 3, no. 5, pp. 16-25, Oct 1996.

[8] N. Haardt and W. Mohr, "The complete solution for third-generation wireless communications: two modes on air, one winning strategy", *IEEE Personal Communications*, vol. 7, no. 6, pp. 18-24, Dec 2000.

[9] K. I. Pedersen, P. E. Mogensen and J. Ramiro-Moreno, "Application and performance of downlink beamforming techniques in UMTS", *IEEE Communications Magazine*, vol. 41, no. 10, pp. 134-143, Oct 2003.

[10] F. Piolini and A. Rolando, "Smart channel-assignment algorithm for SDMA systems", *IEEE Trans. Microwave Theory and Techniques*, vol. 47, no. 6, pp. 693-699, Jun 1999.

[11] G. Tsoulos, J. McGeehan and M. Beach, "Space division multiple access (SDMA) field trials. I. Tracking and BER performance", *IEE Proc. Radar, Sonar and Navigation*, vol. 145, no. 1, pp. 73-78, Feb 1998.

[12] G. Tsoulos, J. McGeehan and M. Beach, "Space division multiple access (SDMA) field trials. 2. Calibration and linearity issues", *IEE Proc. Radar, Sonar and Navigation*, vol. 145, no. 1, pp. 79-84, Feb 1998.

[13] R. L. Peterson, R. E. Ziemer and D. E. Borth, *Introduction to Spread Spectrum Communications*, Upper Saddle River, N.J: Prentice Hall PTR, 1995.

[14] R. Prasad, *CDMA for Wireless Personal Communications*, Artech House Mobile Communications Series, 1996.

[15] P. Varzakas and G. S. Tombras, "Spectral efficiency for a hybrid DS/FH code-division multiple-access system in cellular mobile radio", *IEEE Trans. Vehicular Technology*, vol. 50, no. 6, pp. 1321-1327, Nov 2001.

[16] T. Liu, E. Li and H. Xiang, "A hybrid TH/DS spread spectrum system supporting multirate services", *IEEE Proc. Spread Spectrum Techniques and Applications*, vol. 1, pp. 309-313, Sep 1998.

[17] A. Elhakeem, G. Takhar and S. Gupta, "New code acquisition techniques in spread-spectrum communication", *IEEE Trans. Communications*, vol. 28, no. 2, pp. 249-257, Feb 1980.

[18] P. G. Flikkema, "Spread-spectrum techniques for wireless communication", *IEEE Signal Processing Magazine*, vol. 14, no. 3, pp. 26-36, May 1997.

[19] F. Adachi, M. Sawahashi and H. Suda, "Wideband DS-CDMA for next-generation mobile communications systems", *IEEE Communications Magazine*, vol. 36, no. 9, pp. 56-69, Sep 1998.

[20] H. L. Van Trees, *Optimum Array Processing (Detection, Estimation, and Modulation Theory, Part IV)*, Wiley-Interscience, Mar 2002.

[21] M. S. Bartlett, "Smoothing periodograms from time series with continuous spectra", *Nature*, vol. 161, pp. 686-687, 1948.

[22] J. Capon, "High-resolution frequency-wavenumber spectrum analysis", *IEEE Proc.*, vol. 57, no. 8, pp. 1408-1418, Aug 1969.

[23] R. O. Schmidt, *A signal subspace approach to multiple emitter location and spectral estimation*, PhD thesis, Stanford University, Stanford CA, Nov 1981.

[24] A. Barabell, "Improving the resolution performance of eigenstructure-based direction-finding algorithms", *IEEE Proc. Int. Conf. Acoustic, Speech, and Signal Processing*, vol. 8, pp. 336-339, Apr 1983.

[25] S. V. Schell, R. A. Calabretta, W. A. Gardner, and B. G. Agee, "Cyclic MUSIC algorithms for signal-selective direction estimation", *IEEE Proc. Int. Conf. Acoustic, Speech, and Signal Processing*, vol. 4, pp. 2278-2281, May 1989.

[26] J. Mayhan, L. Niro, "Spatial spectral estimation using multiple beam antennas", *IEEE Trans. Antennas Propagation*, vol. 35, no. 8, pp. 897-906, Aug 1987.

[27] R. Roy and T. Kailath, "ESPRIT - Estimation of signal parameters via rotational invariance techniques", *IEEE Trans. Acoustic, Speech, and Signal Processing*, vol. 37, no. 7, pp. 984-995, Jul 1989.

[28] R. Kumaresan and D. W. Tufts, "Estimating the angles of arrival of multiple plane waves", *IEEE Trans. Aerospace and Electronic Systems*, vol. 19, no. 1, pp. 134-139, Jan 1983.

[29] M. Viberg and B. Ottersten, "Sensor array processing based on subspace fitting", *IEEE Trans. Signal Processing*, vol. 39, no. 5, pp. 1110-1121, May 1991.

[30] P. Stoica and A. Nehorai, "Performance study of conditional and unconditional direction-of-arrival estimation", *IEEE Trans. Acoustic, Speech, and Signal Processing*, vol. 38, no. 10, pp. 1783-1795, Oct 1990.

[31] J. Bohme, "Source-parameter estimation by approximate maximum likelihood and nonlinear regression", *IEEE Journal Oceanic Engineering*, vol. 10, no. 3, pp. 206-212, Jul 1985.

[32] A. G. Jaffer, "Maximum likelihood direction finding of stochastic sources: a separable solution", *IEEE Proc. Int. Conf. Acoustic, Speech, and Signal Processing*, vol. 5, pp. 2893-2896, Apr 1988.

[33] P. Stoica and A. Nehorai, "MUSIC, maximum likelihood, and Cramer-Rao bound", *IEEE Trans. Signal Processing*, vol. 37, no. 5, pp. 720-741, May 1989.

[34] K. C. Sharman, "Maximum likelihood parameter estimation by simulated annealing", *IEEE Proc. Int. Conf. Acoustic, Speech, and Signal Processing*, vol. 5, pp. 2741-2744, Apr 1988.

[35] I. Ziskind and M. Wax, "Maximum likelihood localization of multiple sources by alternating projection", *IEEE Trans. Acoustic, Speech, and Signal Processing*, vol. 36, no. 10, pp. 1553-1560, Oct 1988.

[36] T. S. Rappaport, *Wireless Communications: Principles and Practice*, Upper Saddle River, N.J: Prentice Hall PTR, 1996.

[37] J. G. Proakis, *Digital Communications*, 4th Edition, Boston: McGraw-Hill, 2001.

[38] R. H. Clarke, "A statistical theory of mobile-radio reception", *Bell Systems Tech. J.*, vol. 47, no. 6, pp. 957-1000, Jul-Aug 1968.

[39] W. Lee, "Effects on correlation between two mobile radio base-station antennas", *IEEE Trans. Communications*, vol. 21, no. 11, pp. 1214-1224, Nov 1973.

[40] A. Vigants, "Space-diversity performance as a function of antenna separation", *IEEE Trans. Communications*, vol. 16, no. 6, pp. 831-836, Dec 1968.

[41] P. Beckmann, *The Depolarization of Electromagnetic Waves*, Boulder, Colo: Golem Press, 1968.

[42] G. Porter, "Measurement of polarization statistics of signals received over a short range HF path", *IEEE Trans. Communications*, vol. 14, no. 4, pp. 484-494, Aug 1966.

[43] R. G. Vaughan, "Polarization diversity in mobile communications", *IEEE Trans. Vehicular Technology*, vol. 39, no. 3, pp. 177-186, Aug 1990.

[44] W. Lee and Y. Yeh, "Polarization diversity system for mobile radio", *IEEE Trans. Communications*, vol. 20, no. 5, pp. 912-923, Oct 1972.

[45] S. Kozono, T. Tsuruhara, and M. Sakamoto, "Base station polarization diversity reception for mobile radio", *IEEE Trans. Vehicular Technology*, vol. 33, no. 4, pp. 301-306, Nov 1984.

[46] A. J. Paulraj and E. Lindskog, "Taxonomy of space-time processing for wireless networks", *IEE Proc. Radar, Sonar and Navigation*, vol. 145, no. 1, pp. 25-31, Feb 1998.

[47] C. M. Panazio and F. R. P. Cavalcanti, "Decoupled space-time processing: performance evaluation for a TDMA system", *IEEE Proc. Vehicular Technology Conf.*, vol. 4, pp. 1998-2002, Oct 2001.

[48] T. Abe and K. Araki, "Space-time receivers for CDMA systems", *IEEE Proc. Vehicular Technology Conf.*, vol. 2, pp. 1115-1119, May 2000.

[49] A. F. Naguib, "Space-time receivers for CDMA multipath signals", *IEEE Proc. Int. Conf. Communications*, vol. 1, pp. 304-308, Jun 1997.

[50] B. H. Khalaj, A. Paulraj, and T. Kailath, "2D RAKE Receivers for CDMA Cellular Systems", *IEEE Proc. Global Telecoms Conf.*, vol. 1, pp. 400-404, Dec 1994.

[51] C. Brunner, M. Haardt, and J. A. Nossek, "On space-time RAKE receiver structures for WCDMA", *Proc. 33rd Asilomar Conf. on Signals, Systems, and Computers*, vol. 2, pp. 1546-1551, Oct 1999.

[52] X. Bernstein and A. M. Haimovich, "Space-time processing for increased capacity of wireless CDMA", *IEEE Proc. Int. Conf. Communications*, vol. 1, pp. 597-601, Jun 1996.

[53] A. J. van der Veen, S. Talwar, and A. Paulraj, "A subspace approach to blind space-time signal processing for wireless communication systems", *IEEE Trans. Signal Processing*, vol. 45, no. 1, pp. 173-190, Jan 1997.

[54] F. Pipon, P. Chevalier, P. Vila, and J. J. Monot, "Joint spatial and temporal equalization for channels with ISI and CCI-theoretical and experimental results for a base station reception", *IEEE Proc. Signal Processing Advances in Wireless Communications*, pp. 309-312, Apr 1997.

[55] E. Lindskog, A. Ahlen, and M. Sternad, "Spatio-temporal equalization for multipath environments in mobile radio applications", *IEEE Proc. Vehicular Technology Conf.*, vol. 1, pp. 399-403, Jul 1995.

[56] J. Modestino and M. Eyuboglu, "Integrated multielement receiver structures for spatially distributed interference channels", *IEEE Trans. Information Theory*, vol. 32, no. 2, pp. 195-219, Mar 1986.

[57] R. Kohno, "Spatial and temporal communication theory using adaptive antenna array", *IEEE Personal Communications*, vol. 5, no. 1, pp. 28-35, Feb 1998.

[58] D. A. Pados and S. N. Batalama, "Joint space-time auxiliary-vector filtering for DS/CDMA systems with antenna arrays", *IEEE Trans. Communications*, vol. 47, no. 9, pp. 1406-1415, Sep 1999.

[59] R. O. Schmidt, "Multilinear array manifold interpolation", *IEEE Trans. Acoustic, Speech, and Signal Processing*, vol. 40, no. 4, pp. 857-866, Apr 1992.

[60] J. Li and P. Stoica, "Efficient parameter estimation of partially polarized electromagnetic waves", *IEEE Trans. Signal Processing*, vol. 42, no. 11, pp. 3114-3125, Nov 1994.

[61] K. C. Ho, K. C. Tan and A. Nehorai, "Estimating directions of arrival of completely and incompletely polarized signals with electromagnetic vector sensors", *IEEE Trans. Signal Processing*, vol. 47, no. 10, pp. 2845-2852, Oct 1999.

[62] E. Ferrara, Jr. and T. Parks, "Direction finding with an array of antennas having diverse polarizations", *IEEE Trans. Antennas Propagation*, vol. 31, no. 2, pp. 231-236, Mar 1983.

[63] R. O. Schmidt, "Multiple emitter location and signal parameter estimation", *IEEE Trans. Antennas Propagation*, vol. 34, no. 3, pp. 276-280, Mar 1986.

[64] A. J. Weiss and B. Friedlander, "Performance of diversely polarized antenna arrays for correlated signals", *IEEE Trans. Aerospace and Electronic Systems*, vol. 28, no. 3, pp. 869-879, Jul 1992.

[65] J. Li and R. T. Compton, Jr, "Angle and polarization estimation in a coherent signal environment", *IEEE Trans. Aerospace and Electronic Systems*, vol. 29, no. 3, pp. 706-716, Jul 1993.

[66] J. Li, P. Stoica and D. Zheng, "Efficient direction and polarization estimation with a COLD array", *IEEE Trans. Antennas Propagation*, vol. 44, no. 4, pp. 539-547, Apr 1996.

[67] K. T. Wong, "Direction finding/polarization estimation - dipole and/or loop triad(s)", *IEEE Trans. Aerospace and Electronic Systems*, vol. 37, no. 2, pp. 679-684, Apr 2001.

[68] K. T. Wong and M. D. Zoltowski, "Self-initiating MUSIC-based direction finding & polarization estimation in spatio-polarizational beamspace", *IEEE Trans. Antennas Propagation*, vol. 48, no. 8, pp. 1235-1245, Aug 2000.

[69] K. T. Wong, "Blind beamforming/geolocation for wideband-FFH with unknown hop-sequences", *IEEE Trans. Aerospace and Electronic Systems*, vol. 37, no. 1, pp. 65-76, Jan 2001.

[70] P. H. Chua, C. M. S. See and A. Nehorai, "Vector-sensor array processing for estimating angles and times of arrival of multipath communication signals", *IEEE Proc. Int. Conf. Acoustic, Speech, and Signal Processing*, vol. 6, pp. 3325-3328, May 1998.

[71] L. K. Huang and A. Manikas, "Blind adaptive single-user array receiver for MAI cancellation in multipath fading DS-CDMA channels", *EUSIPCO Proceedings*, vol. 2, pp. 647-650, Sep 2000.

[72] M. C. Vanderveen, C. B. Papadias, and A. Paulraj, "Joint angle and delay estimation (JADE) for multipath signals arriving at an antenna array", *IEEE Communications Lett.*, vol. 1, no. 1, pp. 12-14, Jan 1997.

[73] G. G. Raleigh and T. Boros, "Joint space-time parameter estimation for wireless communication channels", *IEEE Trans. Signal Processing*, vol. 46, no. 5, pp. 1333-1343, May 1998.

[74] Jason W.P. Ng and A. Manikas, "Diversely polarised arrays in DS-CDMA: A space-time channel estimation approach", *IEEE Proc. Int. Conf. Acoustic, Speech, and Signal Processing*, vol. 3, pp. 2617-2620, May 2002.

[75] Jason W.P. Ng and A. Manikas, "Polarisation-Angle-Delay Estimation using crossed-dipole array for DS-CDMA systems", *IEEE Proc. Int. Conf. Digital Signal Processing*, vol. 1, pp. 259-262, July 2002.

[76] A. Manikas and Jason W.P. Ng, "Crossed-dipole arrays for asynchronous DS-CDMA systems", *IEEE Trans. Antennas Propagation*, vol. 52, no. 1, pp. 122-131, Jan 2004.

[77] G. A. Deschamps, "Geometrical representation of the polarization of a plane electromagnetic wave", *Proc. IRE*, vol. 39, pp. 540-544, May 1951.

[78] R. T. Compton, Jr, "The tripole antenna: An adaptive array with full polarization flexibility", *IEEE Trans. Antennas Propagation*, vol. 29, no. 6, pp. 944-952, Nov 1981.

[79] C. A. Balanis, *Antenna Theory: Analysis and Design*, 2nd edition., New York, Chichester: Wiley, 1997.

[80] T. Shan, M. Wax and T. Kailath, "On spatial smoothing for direction-of-arrival estimation of coherent signals", *IEEE Trans. Acoustic, Speech, and Signal Processing*, vol. 33, no. 4, pp. 806-811, Aug 1985.

[81] A. J. Weiss and B. Friedlander, "Maximum likelihood signal estimation for polarization sensitive arrays", *IEEE Trans. Antennas Propagation*, vol. 41, no. 7, pp. 918-925, Jul 1993.

[82] T. P. Jantti, "The influence of extended sources on the theoretical performance of the MUSIC and ESPRIT methods: narrow-band sources", *IEEE Proc. Int. Conf. Acoustic, Speech, and Signal Processing*, vol. 2, pp. 429-432, Mar 1992.

[83] K. I. Pedersen, P. E. Mogensen and B. H. Fleury, "Spatial channel characteristics in outdoor environments and their impact on BS antenna system performance", *IEEE Proc. Vehicular Technology Conf.*, vol. 2, pp. 719-723, May 1998.

[84] M. Larsson, "Spatio-temporal channel measurements at 1800 MHz for adaptive antennas", *IEEE Proc. Vehicular Technology Conf.*, vol. 1, pp. 376-380, May 1999.

[85] A. Kuchar, J. P Rossi and E. Bonek, "Directional macro-cell channel characterization from urban measurements", *IEEE Trans. Antennas Propagation*, vol. 48, no. 2, pp. 137-146, Feb 2000.

[86] Y. Oda and T. Taga, "Clustering of local scattered multipath components in urban mobile environments", *IEEE Proc. Vehicular Technology Conf.*, vol. 1, pp. 11-15, May 2002.

[87] Y. Meng, P. Stoica and K. M. Wong, "Estimation of the directions of arrival of spatially dispersed signals in array processing", *IEE Proc. Radar, Sonar and Navigation*, vol. 143, no. 1, pp. 1-9, Feb 1996.

[88] S. Valaee, B. Champagne and P. Kabal, "Parametric localization of distributed sources", *IEEE Trans. Signal Processing*, vol. 43, no. 9, pp. 2144-2153, Sep 1995.

[89] Q. Wu, K. M. Wong, Y. Meng and W. Read, "DOA estimation of point and scattered sources - Vec-Music", *IEEE SP Workshop on Statistical Signal and Array Processing*, pp. 365-368, Jun 1994.

[90] T. Trump and B. Ottersten, "Estimation of nominal direction of arrival and angular spread using an array of sensors", *Proc. Signal Processing*, vol. 50, pp. 57-69, Apr 1996.

[91] Q. Wan, J. Yuan, Y. N. Peng, W. L. Yang and K. P. Long, "Estimation of nominal direction of arrival and angular spread using the determinant of the data matrix", *Int. Workshop on Mobile and Wireless Communications Network*, pp. 76-79, Sep 2002.

[92] M. Bengtsson and B. Ottersten, "Low-complexity estimators for distributed sources", *IEEE Trans. Signal Processing*, vol. 48, no. 8, pp. 2185-2194, Aug 2000.

[93] O. Besson and P. Stoica, "Decoupled estimation of DOA and angular spread for a spatially distributed source", *IEEE Trans. Signal Processing*, vol. 48, no. 7, pp. 1872-1882, Jul 2000.

[94] P. D. Karaminas and A. Manikas, "Superresolution broad null beamforming for co-channel interference cancellation in mobile radio networks", *IEEE Trans. Vehicular Technology*, vol. 49, no. 3, pp. 689-697, May 2000.

[95] Jason W.P. Ng and A. Manikas, "Diffused channel framework for blind space-time DS-CDMA receiver", *Hellenic European Research on Computer Mathematics & its Applications (HERCMA) - Special Topics in Communications*, vol. 1, pp. 183-187, Sep 2003.

[96] Jason W.P. Ng and A. Manikas, "Blind Diffused Space-Time DS-CDMA Receiver", *International Journal HERMIS-mu-pi (Mathematics & Informatics Science)*, vol. 4, pp. 131-152, 2003.

[97] D. Asztely, B. Ottersten and A. L. Swindlehurst, "Generalised array manifold model for wireless communication channels with local scattering", *IEE Proc. Radar, Sonar and Navigation*, vol. 145, no. 1, pp. 51-57, Feb 1998.

[98] D. Asztely and B. Ottersten, "The effects of local scattering on direction of arrival estimation with MUSIC", *IEEE Trans. Signal Processing*, vol. 47, no. 12, pp. 3220-3234, Dec 1999.

[99] T. S. Naveendra and A. Manikas, "Spatio-temporal Array (STAR) DS-CDMA systems: Localized scattering channel estimation", *IEEE Proc. Global Telecoms Conf.*, Nov 2002.

[100] J. C. Liberti and T. S. Rappaport, *Smart Antennas for Wireless Communications: IS-95 and Third Generation CDMA Applications*, Upper Saddle River, N.J: Prentice Hall PTR, 1999.

[101] D. H. Johnson and D. E. Dudgeon, *Array Signal Processing: Concepts and Techniques*, Englewood Cliffs: PTR Prentice Hall, 1993.

[102] S. M. Alamouti, "A simple transmit diversity technique for wireless communications", *IEEE Selected Areas in Communications*, vol. 16, no. 8, pp. 1451-1458, Oct 1998.

[103] G. J. Foschini, "Layered space-time architecture for wireless communication in a fading environment when using multiple antennas", *Bell Labs Technical Journal*, vol. 1, no. 2, pp. 41-59,1996.

[104] H. Huang, H. Viswanathan and G. J. Foschini, "Multiple antennas in cellular CDMA systems: transmission, detection, and spectral efficiency", *IEEE Trans. Wireless Communications*, vol. 1, no. 3, pp. 383-392, Jul 2002.

[105] R. T. Derryberry, S. D. Gray, D. M. Ionescu, G. Mandyam and B. Raghothaman, "Transmit diversity in 3G CDMA systems", *IEEE Communications Magazine*, vol. 40, no. 4, pp. 68-75, Apr 2002.

[106] B. Hochwald, T. L. Marzetta, C. B. Papadias, “A transmitter diversity scheme for wideband CDMA systems based on space-time spreading”, *IEEE Selected Areas in Communications*, vol. 19, no. 1, pp. 48-60, Jan 2001.

[107] J. C. Liberti and T. S. Rappaport, *Smart Antennas for Wireless Communications: IS-95 and Third Generation CDMA Applications*, Upper Saddle River, N.J: Prentice Hall PTR, 1999.

[108] L. C. Godara, “Application of antenna arrays to mobile communications, Part I: Performance improvement, feasibility, and system considerations”, *IEEE Proc.*, vol. 85, no. 7, pp. 1031-1060, Jul 1997.

[109] L. C. Godara, “Application of antenna arrays to mobile communications, Part II: Beam-forming and direction-of-arrival considerations”, *IEEE Proc.*, vol. 85, no. 8, pp. 1195-1245, Aug 1997.

[110] H. Krim and M. Viberg, “Two decades of array signal processing research: The parametric approach”, *IEEE Signal Processing Magazine*, vol. 13, no. 4, pp. 67-94, Jul 1996.

[111] A. J. Paulraj and C. B. Papadias, “Space-time processing for wireless communications”, *IEEE Signal Processing Magazine*, vol. 14, no. 6, pp. 49-83, Nov 1997.

[112] Jason W.P. Ng and A. Manikas, "MIMO Array DS-CDMA System: A Blind Space-Time-Doppler Estimation/Reception", *IEEE Proc. Int. Conf. Acoustic, Speech, and Signal Processing*, vol. 2, pp. 337-340, May 2004.

[113] A. M. Sayeed and B. Aazhang, “Joint multipath-Doppler diversity in mobile wireless communications”, *IEEE Trans. Communications*, vol. 47, no. 1, pp. 123-132, Jan 1999.

[114] E. N. Onggosanusi, A. M. Sayeed and B. D. Van Veen, “Canonical space-time processing for wireless communications”, *IEEE Trans. Communications*, vol. 48, no. 10, pp. 1669-1680, Oct 2000.

[115] I. F. Akyildiz, W. Su, Y. Sankarasubramaniam and E. Cayirci, “A survey on sensor networks”, *IEEE Communications Magazine*, vol. 40, no. 8, pp. 102-114, Aug 2002.

[116] K. Rohani and L. Jalloul, "Orthogonal Transmit Diversity for direct spread CDMA", *Motorola contribution to ETSI SMG2 UMTS-L1*, Stockholm Sweden, Sep 15-17 1997.

[117] 3GPP2, *Physical Layer Standard for cdma2000 Spread Spectrum Systems Release D*, Document no. C.S0002-D v1.0, Feb 2004.

[118] R. Bansal, "Tripole to the rescue", *IEEE Antennas Propagation Magazine*, vol. 43, no. 2, pp. 106-107, Apr 2001.

[119] E. N. Onggosanusi, B. D. Van Veen, and A.M. Sayed, "Spatio-temporal polarization signaling for multipath spread-spectrum channels", *IEEE Proc. Int. Conf. Acoustic, Speech, and Signal Processing*, vol. 5, pp. 2837-2840 , Jun 2000.

[120] G. F. Hatke, "Performance analysis of the SuperCART antenna array", MIT Lincoln Lab., Lexington, MA, Project Report AST-22, Mar 1992.

[121] J. F. Bull and L. R. Burgess, "A compact antenna array for direction finding in the HF band", *Proc. Tactical Communications Conf.*, vol. 1, pp. 651-657, Apr 1990.

[122] A. Nehorai and E. Paldi, "Vector-sensor array processing for electromagnetic source localization", *IEEE Trans. Acoustic, Speech, and Signal Processing*, vol. 42, no. 2, pp. 376-398, Feb 1994.

[123] C. M. S. See and A. B. Gershman, "Subspace-based direction finding in partly calibrated arrays of arbitrary geometry", *IEEE Proc. Int. Conf. Acoustic, Speech, and Signal Processing*, vol. 3, pp. 3013-3016, May 2002.

[124] Jason W.P. Ng, T. Zhang and A. Manikas, "Space-Time Block Coding Based MIMO Array Receiver", *Int. Symposium on Wireless Communication Systems*, Sep 2004.

[125] V. Le Nir, M. Helard, and R. Le Gouable, "Space-time block coding applied to turbo coded multicarrier CDMA", *IEEE Proc. Vehicular Technology Conf.*, vol. 1, pp. 577-581, Apr 2003.

[126] E. H. Callaway, *Wireless Sensor Networks: Architectures and Protocols*, Boca Raton: CRC Press, 2003.

[127] J. C. Chen, Y. Kung, and R. E. Hudson, "Source localization and beamforming", *IEEE Signal Processing Magazine*, vol. 19, no. 2, pp. 30-39, Mar 2002.

[128] L. Cong and W. Zhuang, "Hybrid TDOA/AOA mobile user location for wideband CDMA cellular systems", *IEEE Trans. Wireless Communications*, vol. 1, no. 3, pp. 439-447, Jul 2002.

Index